THE DDC,

THE UNIVERSE OF KNOWLEDGE,

AND

THE POST-MODERN LIBRARY

THE DDC,

THE UNIVERSE OF KNOWLEDGE,

AND

THE POST-MODERN LIBRARY

by

Francis L. Miksa

FOREST PRESS
A Division of
OCLC Online Computer Library Center, Inc.
ALBANY, NEW YORK
1998

Library of Congress Cataloging-in-Publication Data

Miksa, Francis L., 1938–
    The DDC, the universe of knowledge, and the post-modern library / by Francis Miksa.
      p.   cm.
    Includes bibliographic references.
    ISBN 0-910608-64-4 (alk. paper)
    1. Classification, Dewey decimal--History. 2. Classification of sciences--History--19th century. 3. Classification of sciences--History--20th century. I. Title.
Z696.D7M63 1997   97-41719
025.4´31´09--dc21   CIP

Recycled Paper

CONTENTS

ACKNOWLEDGMENTS

This account of the DDC and ideas related to it represents a much expanded version of a paper under the same title given as the DDC Anniversary Lecture at the 4th International ISKO Conference 15–18 July 1996 at the Library of Congress, Washington, D.C. An extended abstract of the paper may be found in *Knowledge Organization and Change: Proceedings of the Fourth International ISKO Conference, 15–18 July 1996, Washington, D.C., U.S.A.*, edited by Rebecca Green, in *Advances in Knowledge Organization* 5 (1996): 406–12 (Frankfurt/Main: Indeks Verlag).

The author wishes to acknowledge his enormous gratitude to OCLC Forest Press for research support basic to this work and, especially, to Peter J. Paulson, executive director of Forest Press, for his general encouragement and patience in all aspects of this endeavor. I should also like to extend my thanks and appreciation to Forest Press staff: to Judith Kramer-Greene for reading the manuscript so thoroughly, for making many invaluable editorial suggestions, and for supervising the production of the book; to Elizabeth Hansen for her excellent copyediting and proofreading comments; and to Judith Pisarski for her careful typing of the manuscript. Needless to say, the opinions found in the text are the author's alone.

INTRODUCTION

Melvil Dewey originally created his "Decimal Classification" (DDC) while a student and librarian at Amherst College between 1872 and 1876. He published it at the first meeting of the American Library Association in Philadelphia at the United States Centennial Exhibition in 1876. In his book on the first 18 editions of the DDC, John P. Comaromi, following Leo LaMontagne in part, concludes that its lineage is notable. In broad brush strokes he traces its main class structure back four or five years to the work of William Torrey Harris at the St. Louis Public School Library; from Harris back through three more decades to the work of Edward Johnston at three different institutions—the College of South Carolina, the Mercantile Library of New York City, and the Mercantile Library of St. Louis; and from Johnston back an additional two centuries to the work of Francis Bacon. Comaromi also agrees in part with the claims of Eugene Graziano that, through Harris, the Decimal Classification main class arrangement was at least partially influenced by Hegelianism. Comaromi also believes that the claims of John Maass have some possible merit. Maass claimed that Dewey may have borrowed the idea of using a decimal notation to number areas of knowledge from William Phipps Blake, who had used it as early as 1872 for arranging the exhibits at the then upcoming Centennial Exhibition in Philadelphia (Comaromi 1976, pp. 1–29; LaMontagne 1961, pp. 173–88).

Comaromi's discussion of the broad lineage of the scheme deals mainly with its superstructure—the origin and order of its ten main divisions—and, briefly, with the origin of its notation. As for the order within the system—for example, at the second or third summary levels of the 1st edition—Comaromi confines himself chiefly to describing the resultant scheme, with occasional speculations as to why one or another subject sequence had been adopted. He also notes that Dewey had sent copies of his work to various librarian experts for review. But no information exists as to any responses Dewey might have received, although in the introduction to the 1st edition Dewey acknowledges the "valuable suggestions and appreciative criticism" of Charles A. Cutter of the Boston Athenaeum and John Fiske of the Harvard University Library (DDC 1st ed., p. 10). With the exception of attributing the order of the second hierarchical level of the 500s (Sciences) and the order of particular classes within the sciences to Alexander Bain's textbook, *Logic: Deductive and Inductive,* Comaromi does not discuss the origins of the internal order of the system other than to repeat Dewey's own hint that choices of topics and their order within the various divisions were greatly influenced by Amherst College professors (Comaromi 1976, pp. 59–68; DDC 1st ed., pp. 10–11).

More recently, Wayne Wiegand has extended our knowledge of the origins of the scheme several steps further by showing its relationship to at least one textbook Dewey used while a student at Amherst—Denison Olmstead's *Introduction to Natural Philosophy* (2d ed., 1870). This work had been revised by Ebenezer Snell, an Amherst professor, and was used at the school. Wiegand also demonstrates the probable influence on the scheme of three specific Amherst faculty members, Edward Hitchcock, Edward Seelye, and John W. Burgess. However, Wiegand has been able to assemble only circumstantial evidence as to which subject areas were directly affected, and this mainly in the form of themes found in the professors' teaching and writing (Wiegand 1996b).

As useful as the previous discussions are in attempting to understand the origins of this most notable of all modern general classification schemes, none of them accounts fully for the unique phenomenon of the system. None of them answers the much more trenchant question of why someone like Melvil Dewey—it could have been anyone, but here we must contend with Dewey himself—would indulge in this kind of exercise at all. Why, when attempting to create a practical system to make it easier to find books on particular topics, would one construct a system of classes? Why would one construct a system which is hierarchically arranged in the form of ten main divisions of what is purported to be all knowledge, each main division further subdivided in up to nine subdivisions, and those, in turn, subdivided in up to nine more subsubdivisions, all of which carry the implication of "logical" order? Why not construct some other device—for example, an open-ended set of alphabetically arranged main classes under which new subclasses could always be inserted? Or, why not create a catalog of alphabetically arranged topics found in books, organize the books on the shelves in some compact way, and provide a sequential inventory notation to connect the two?[1]

Considering the fact that typical libraries of the time were quite small, possibly an average of only two to five thousand volumes each, and that even exemplar libraries were not the size we know today, it may well seem that Dewey's solution to a practical need was clearly a case of overkill. Certainly, it had something of a pretentious cast. Just think of it! How could anyone presume to organize books logically and systematically for all knowledge? This conjures up images of Francis Bacon devising a *Novum Organum* to replace Aristotle's *Organon*. There can be little doubt that Dewey fully intended to make an extensive, hierarchically arranged scheme which covered all knowledge, a scheme that could be used in libraries to organize books according to the knowledge the books contained. Pretentious? Surely!

I insist on the question—"Why this form of such a tool?"—for two reasons. First, the question has been raised by many in our own field of library and information science. For example, since the 1950s the generation of librarians who have become accustomed to computerized natural language indexing systems have questioned why such an investment is made in a system like the DDC (or for

---

[1] These two suggestions are represented: (1) by an alphabetico-classed catalog, such as the one created at Harvard College by Ezra Abbott during the 1860s. In this catalog, a series of main classes was arranged alphabetically and was then subdivided by subclasses. These subclasses were also arranged alphabetically under their respective heads, and were, in turn, subdivided by subclasses, also arranged alphabetically under their respective heads, and so on; and (2) by a simple dictionary catalog, of the kind popularized by William S. Poole from the 1850s on, or even its more sophisticated syndetic variant created by C. A. Cutter in the 1870s.

that matter in any other general system regardless of its structural principles). Computerized indexing systems seem to be a more efficient device for finding documents on a subject. Criticism has also emanated from the broader world as well. Witness Steve G. Steinberg's 1996 conclusion in *Wired* that the DDC is essentially old-fashioned—poor at dealing with new fields of knowledge, limited by being a "one-place" shelf classification, and tied to an outmoded view of the world.[2]

Second, I insist on the question because of the implications of the responses I have received over the years from many practicing American librarians and library students. With only a few exceptions, practicing librarians and students alike often seem to be at a loss when confronted with the dual nature of the system, with its ability to provide both an inventory control system (sometimes called its "mark and park" function), and a system to classify books. In most instances, practitioners find the first goal to be the system's chief purpose, and the second to be mysterious, at best.

In the end, we are left with the question I asked originally—Why did Dewey pick this particular method and not some other? If this question is applicable to the DDC, it is applicable as well for all other systematic library classification efforts. Why have librarians worked so hard at keeping the DDC (or an alternative) going? Why have they invested so much labor and money in applying it (and other similar schemes)? Why have they persisted in believing that a systematic, hierarchical classificatory structure such as the DDC (or some alternative) is necessary?

I am not sure that a completely satisfying answer can be made to these questions, but I wish at least to try To do this I will direct my remarks to four areas. First, I will summarize the long history of the DDC in order to provide an overview of the system and an explanation of how it has arrived at its present state. Second, I will place the DDC in its historical context in the nineteenth and early twentieth centuries in order to explain how it was originally conceived, what it was and was not intended to be. Third, I will address how the DDC reflects developments since that formative period, especially since the 1920s, in order to demonstrate how it has accommodated the growth of library classification theory. Last, I will suggest by way of summary and conclusions an alternative approach to the use of the DDC (and to library classification in general) as our field enters what I think is aptly called a "post-modern" age.

---

[2] Steinberg writes that the DDC is "Unsurprisingly . . . poor at classifying knowledge in 'newly' established fields like genetics or electrical engineering." Later in the same article, Steinberg more or less dismisses the DDC as an expression of "the worldview . . . of a 19th-century Puritan named Melvil Dewey." He concludes that its nineteenth-century origins causes it to be "constantly struggling against obsolescence" (Steinberg 1996, pp. 109-110, 180). It might be of interest to Steinberg that Dewey began using the category "Electrical engineering" as early as the 3rd edition of the DDC in 1888. As to Dewey being a Puritan, nothing could be farther from the truth. His charge that the scheme had a nineteenth-century world view is accurate only of its editions produced in the nineteenth century. It has retained some of its oldest collocations, but only where the collocations are still widely held by society. In other places, such collocations have long since been abandoned. The fact is that the scheme has changed enormously with the times and Steinberg simply knows very little of it. He seems to have gotten his information about the scheme from second-hand sources and has not taken the time to examine the system for himself.

PART I

Decimal Classification Developments

The development of the DDC over its long history can be usefully divided into the following sections: Beginnings—Editions 1–6 (1876–1899); Conflict—Editions 7–15 (1911–1953); Recovery and Advance—Editions 16–21 (1958–1996).[1]

Beginnings: Editions 1–6 (1876–1899)

The first six editions of the DDC constitute the period of its beginnings and extend from the founding of the system to its wide acceptance in the library community. More specifically, editions one through three (1876, 1885, and 1888) correspond to its creation and expansion into a full-blown library classification system while librarians were first experimenting with it and debating the merits of the "close classification" that it represented.[2] In contrast, editions four through six (1891, 1894, and 1899) represent something of a plateau in its development as a system with little new material added to it except index entries. These three editions function chiefly as new printings of a system which by the 1890s had become widely accepted in the library community and were in considerable demand.

Editions 1–3 (1876–1888)

Historical interest in the 1st edition of the DDC has generally focused on issues related to its origins and the order of its main classes. Many have also noted

---

[1] The chief sources for this brief history are Comaromi (1976), my close examination of each of the editions, and correspondence related to the DDC for the period 1921–1932 found in the Melvil Dewey Papers, Columbia University Library. These sources will be cited hereafter only for quotations or other less obvious kinds of information. Sources other than these will be cited normally.

[2] "Close classification" refers to a classification system in which subclasses were extended downward to considerable hierarchical levels. It was the practice of many classification systems before Dewey's time to subdivide such schemes sparingly so that when a book was assigned to a particular position in the hierarchy, that actual position might not match in scope the subject being indicated. For example, a book on Dogs might actually be placed in a category such as Mammals rather than the specific category Dogs. To classify this way—typically called "broad classification"—was to provide "class entry," where the category assigned did not represent the particular subject being indicated but rather the class of which the specific category was a part. All library classification schemes tend to practice class entry some of the time. However, before Dewey invented his scheme and, in fact, before the second edition of his system, class entry tended to be the rule rather than an occasional exception. In the 1880s, some librarians considered the practice of extending classification schemes downward in this way in order to accommodate particular subjects of books both confusing and a waste of effort and, therefore, argued against the practice. Dewey, along with Charles A. Cutter, argued against this objection claiming that a scheme of categories had to be developed in this way in order to accommodate growing library collections. It should be noted that the issue of close classification as described here was different than the twentieth-century argument originating with Ranganathan that a classification system must be able to specify the entire subject content of a book with a single heading which was co-extensive with the topical content of the book (i.e., that matched the scope of a document). Both arguments are discussed in Miksa (1983a).

how small it was—merely forty-four pages for all of its parts—and quickly go on to describe later editions as more representative of what the DDC has become. Assigning the 1st edition to the category of "puny" does not do it justice, however, for in many respects it contained much of what the system became in another decade. In fact, there are three remarkable features of it that deserve more attention than they have received.

First, this initial edition of the DDC contains many more unique subject categories than simply the total number of class numbers in its twelve pages of schedules. This is initially apparent if one merely looks at the occurrence of multiple term headings that accompany notations in the schedules. The 921 unique numbers of the schedules contain not simply 921 terms, but rather 1,044 terms, the 123 extra terms of the larger count indicative of subtopics which did not have their own unique numbers. This occurred in many places but most often at the third hierarchical level.

For example, class 240 at the second hierarchical level in the schedules was given the phrase PRACTICAL AND DEVOTIONAL to indicate two separate categories of THEOLOGY (class 200). The first was PRACTICAL THEOLOGY, the second was DEVOTIONAL THEOLOGY. The phrases are not synonymous and both are found in the index of the scheme. Both refer to the 240 number in the schedules but neither is further specified in the schedules with a third hierarchical number. Likewise, at the third hierarchical level, the notation 183, a division of ANCIENT PHILOSOPHIES (class 180), is given the phrase SOPHISTIC AND SOCRATIC by which is meant SOPHISTIC PHILOSOPHY on the one hand and SOCRATIC PHILOSOPHY on the other hand. These constitute two separate categories that are intermixed and not given their own separate numbers. To have done so would have required a fourth hierarchical level in the scheme. Both, however, are listed in the index.

A larger total number of unique subject categories also comes to light when the Relative Index is examined closely. In the index one will find, much as expected, index terms which point exactly to the categories in the schedules, as well as terms representing syntax variations and synonyms for them. For example, the index contains the following two entries: PHYSICAL GEOGRAPHY and GEOGRAPHY, PHYSICAL. Both entries refer to 551, where the schedule lists the term PHYSICAL GEOGRAPHY. In short, one of the terms in the index is the same as the schedule entry, the other represents a synonymous term that varies only in syntax.

Remarkably, however, the index also contains many terms that point to subcategories that did not have a unique notation or listing in the 1st edition. For example, the index term RIVERS also refers to 551 in the schedule. But the term RIVERS (i.e., a particular geographic feature of the category PHYSICAL GEOGRAPHY)[3] is not listed in the schedule. In order to see how extensive the practice of including such additional entries was, I tested several sections. The result was that no less than sixty-four percent of the index terms in these tested sections (not

[3] Category 551 actually contains two categories, PHYSICAL GEOGRAPHY and METEOROLOGY. This means that books on both topics are to be placed at this number. However, one will also find index entries pointing to 551 for BAROMETER, RAIN, STORMS, and TEMPERATURE, all of which clearly constitute subcategories under METEOROLOGY.

including synonyms to these terms) did not have their own unique number. If this is true of the entire index, then the total number of unique categories contained in the system is not 921, or even 1,044, but rather in excess of 2,500. Stated otherwise, there are many more terms in the index standing outside the notational structure of the system than there are inside it.[4] This feature alone made even the "puny" 1st edition quite possibly one of the largest library classification systems ever devised.

The second remarkable feature of the 1st edition is that even here Dewey had already taken important steps towards the systematization of form, genre, language, and geographical categories which were essential to the classification of books as books, but non-essential or "accidental" to the classification of pure knowledge categories. For example, Dewey had established general form categories which he then applied to the first part of each main class (i.e., to the sections 100–109, 200–209, etc.), philological subcategories which he then enumerated in each of the language classes from 410 to 489 (i.e., 411, 412, etc., 451, 452, etc.), and literary genre subcategories which he then enumerated in each of the literature sections from 810 to 889 (i.e., as 811, 812, 813, etc., 851, 852, 853, etc.). In addition, he also established in a single place in the schedules—550s Geology—the principle of using a geographical or place facet to be synthesized with topical headings. He did so under the guise of a "Divide like" instruction, however, rather than in terms of an additive device. In creating all of these provisions, one sees the beginnings of what would eventually become the faceted and synthetic side of the system.

The third remarkable feature of the system is simply the way Dewey spoke about it in his introduction—that is, as an open and expanding system. Dewey spoke of further expansion of the system by the addition of fourth and even fifth digits where appropriate, explaining by way of example that the Amherst library had already extended history notations for countries by including a fourth digit for time periods. To the latter he added, "Should the growth of any of these subsections warrant it, a fifth figure will be added, for the scheme admits of expansion without limit" (DDC 1st ed., p. 5). In sum, the enormous number of categories in the system (both those that were listed in the schedules and those that were hidden in the index), the system's structural properties that are related to what we now call faceted structure and synthesis, and his statements about how the system might be extended, together suggest strongly that Dewey not only fully intended to make this already notably large system even larger, but aimed to do so in a way that was extremely open-ended. That this vision of a large open-ended system

---

[4] The test involved checking index entries which pointed to eight of the ninety-five "tens" sections of the system—the 030s, 130s, 230s, 330s. 530s, 630s, 730s, and 820s. The test was of the variety called "quick and dirty," with reasonable, although not systematic, care taken to differentiate equivalency terms from others. Still, even if subsequent comparisons prove the test to be off the mark to some degree, I am confident that it will not be off by much. For example, even were index entries pointing to non-listed subcategories to amount to only 50%, the total unique categories in the system would still be in excess of 2200. The figures in this test were as follows. 1) Total terms in the index, 2612. Total terms pointing to the sections tested, 343. Distribution of the latter: Pointers exactly to schedule terms, 86; Equivalent terms to the latter, 32; Total remaining terms, 225; Total equivalency terms to the latter, 15; Total terms representing unique subcategories, 220 or 64% of 343.

indeed became the case proved true in the next decade with the 2nd and 3rd editions of the system.

The 2nd and 3rd editions of the DDC (1885, 1888) served chiefly to expand the DDC to a full-blown library classification system. A gross measure of this expansion can be seen in the sheer number of pages in these two editions—314 and 416, respectively—although this is somewhat deceptive because the 3rd edition did not have the forty-eight introductory pages of the second. More specific features included:

1) Filling out all of the previously unused second hierarchical level sections of the scheme (070s, 080s, 090s, and 120s) as well as a number of third level sections (114–119, 164–169), incorporating as fourth (or greater) hierarchical level notations many subdivisions previously listed only in the index, and in general expanding all sections of the scheme often to fourth and fifth levels of hierarchy.

2) Expanding the 800s enormously by listing literary authors by their names under the various time periods of each genre within each national literature and by placing these names in the index.

3) Revising the order of the general form divisions slightly, changing their notations to the familiar 01–09 (numbers which would remain standard until the 13th edition in 1932), increasing the number of terms that were included in the nine categories, supplying a special alphabetical index to the terms beginning in the 3rd edition, and enumerating the divisions in the schedules at many second hierarchical level locations and even at one third hierarchical level location.

4) Expanding philological subcategories to fifty-one in the 2nd edition and to fifty-five in the 3rd edition, placing them in a special table with instructions as to how they should be synthesized with the notations for individual languages, and supplying a special index of those categories which by the 3rd edition had 159 entries.

5) Supplying a special alphabetical list of languages and their notations to be used in both the 400s (Philology) and Literature (800s), the total number of languages going from eighty in the 2nd edition to ninety-five in the 3rd, with the total number of entries reaching 133 in the 3rd. Together, the philological subcategories list and the languages list amounted to a "mini" faceted classification within the system.

6) Expanding geographical "divide-like" provisions from one place in the 1st edition, to 50 locations in the 2nd edition and to 112 locations in the third, with a special alphabetical table of those locations in the schedules, the latter including cross-references.

7) Including at least one instance of a global "divide-like"—i.e., "divide like 100–999"—important as a harbinger of the kind of subject collocation which would become possible in the UDC and in later editions of the DDC.

Most of the expansion of the schedules took place in the 2nd edition, but with those expansions Dewey also found it necessary to make other changes that were disconcerting to early adopters of the system. One such change consisted of the decision to make all notations no less than three digits, even where they were shorter than that in a decimal sense. He did this by extending any shorter notation to three digits by using extra zeros. Thus, Library economy would be 020, not simply 02, and Science would be 500, not simply 5.[5]

A more significant kind of change consisted of rearranging various sections and relocating various individual subjects. As a result, the meanings of many notations changed. Many such changes involved only single notational categories scattered throughout the schedules. In two or three places, however, changes were quite extensive—for example, moving the contents of the 020s (Collections) to the 080s to make a place for Library economy, and creating an American literature section in the 810s. Dewey moved the previous content of the 810s (Treatises and Collections) to the 808 section for literature in general and to the 08 General form divisions position which had previously been empty. He also adjusted the genre heading list under each literature by combining Satire and Humor at xx7, placing Miscellany at xx8, and leaving xx9 open.[6]

Dewey explained in the introductory section that he regretted making so many changes and that to ease the problem of accommodating them, a conversion table listing all such changes had been supplied. He also explained by way of justification that he had been careful to make only those changes which were absolutely necessary. As a way to alleviate suspicions that this might become a regular practice in future editions, he assured his readers that henceforth he would adhere to a strict policy of not changing the basic meanings of notations unless it was absolutely critical to do so. This, of course, was his policy of "integrity of numbers" which was to become so important in later years.

Editions 4–6 (1891–1899)

The 3rd edition marked the temporary end of the initial expansion of the DDC to a full-blown library classification system. The system had, in effect, reached a plateau in its development which would not be advanced upon substantially until the second period of its development beginning with the 7th edition in 1911. The 4th edition in 1891 was larger than the 3rd, but this was due chiefly to the restoration of an introduction to the system which had been omitted from the 3rd edition, and to the slight expansion of the Relative Index. The 5th edition (1894) differed from the fourth by including only an additional page of material, and the 6th edition

---

[5] It should be noted that even with this change, one can find ample evidence that even Dewey sometimes confused the dot after the third digit with a decimal point when, in actuality, the decimal point in the system is assumed and hidden, technically appearing immediately to the left of any notation.

[6] The "xx" in xx7, xx8 and xx9 refers to the beginning digits of any literature number. For example, American literature was found at the 810s. Its beginning digits of 81 when put in place of the "xx" in xx7 would yield 817 (American literary satire and humor) and when put in place of the "xx" in xx8 would yield 818 (American literature miscellany), with 819 being left open.

(1899) differed from the 4th and 5th editions chiefly by expanding the index and by adding period subdivisions to 973 (American history, in general).

What did change in a notable way for the DDC during the 1890s was the environment surrounding the system. First, May Seymour became the editor of the system with the 4th edition, a change which indicated Dewey's personal movement away from day-to-day work on the system itself toward a more executive management role. Seymour had been a student in Dewey's School of Library Economy when it first opened in 1887 and brought to her work, besides considerable editorial skills and an expertise in the area of education, a strong loyalty to Dewey himself. She worked closely with Walter S. Biscoe on the scheme. Biscoe had been Dewey's first assistant at both Amherst and Columbia and was instrumental in bringing out the 2nd and 3rd editions. Second, work on the scheme moved to the New York State Library in Albany, New York, where Dewey found himself so deeply involved in a variety of professional activities that his attention was diverted from any serious further development of the scheme.

Third, and by far the most important factor affecting the future of the scheme, is that by the 1890s the DDC had reached a point of wide acceptance among all kinds of libraries. As such, it became accepted in the library community as a standard tool. However, with that acceptance and use, the DDC also came under closer scrutiny than when it was still being developed during the previous decade. Controversies had arisen earlier, but most of them were on the general nature of library classification systems. In the 1890s a new kind of criticism appeared. It arose most often from users of the system and more often than not consisted of specific evaluations of provisions of the scheme that affected particular libraries. Within this context, the most significant criticisms were those which raised questions about its use in libraries serving specialists and, within that context, the degree to which the scheme failed to be either very logical or very scientific.

Finally, it was in this period as well that two particularly significant events took place which were to affect the future of the system. First, after suitable negotiations in 1895, Dewey gave his consent for the scheme to be used by Paul Otlet and Henri La Fontaine as the basis for the classified catalog of the newly created Institut Internationale de Bibliographie (IIB) in Brussels. Dewey also agreed to cooperate with this initiative in the future development of the system. Second, in contrast to the Belgian initiative and after what turned out to be unsatisfactory negotiations, Dewey refused to allow the Library of Congress to use the system as a basis for the needed reclassification of its own collections. He did so because Library of Congress personnel wanted the freedom to revise the system extensively, and Dewey would not allow this.[7]

Conflict: Editions 7–15 (1911–1952)

The second period in DDC development extended over a half-century and is notable for conflict related to its purposes and objectives. The participants in the

---

[7] The reasons why this failure occurred are complex and can be reviewed both in Comaromi (1976) and in Scott (1970). It suffices to say here only that it leaves all of us with a continuing case of pondering "What if . . ." had the situation turned out differently.

conflict included members of the DDC editorial and administrative staff, librarians of diverse backgrounds, the American Library Association (ALA), the Lake Placid Club Education Foundation, and those responsible for the Classification Décimale of the IIB. Besides personal issues of authority and loyalty, the conflict turned on a variety of interrelated issues: the size of libraries that the system should serve (medium or large), the subject orientation of the libraries that the system should serve (general or special), the bibliographical purpose of the system (shelf classification or catalog access), and a variety of opinions as to the degree of theoretical or scientific accuracy the system should have. Each of the editions of the DDC in this era expressed something of these issues, but for expediency, they will be treated in two larger groups: the 7th to 11th editions on the one hand, and the 12th to 15th editions on the other.

Editions 7–11 (1911–1921)

The 7th edition of the DDC (1911) was published fully a dozen years after the 6th (1899). It had been delayed in many respects by other duties that had fallen to May Seymour and Walter S. Biscoe, and by Melvil Dewey's own professional struggles and conflicts. But after Dewey withdrew from active involvement in library matters in 1906 and had moved the DDC editorial offices to Lake Placid, New York, work resumed on the system in earnest, and this led to the publication of the 7th edition in 1911. Thereafter, four more editions (8th–11th, 1913–1922) were published in the relatively short span of 11 years.

Even before work had resumed on the 7th edition midway through the first decade of the century, a chorus of voices had arisen debating the issue of whether a good library classification should have a practical or a theoretical orientation. A practical classification was considered to be one that was only as extensive in size as was needed by libraries. It would accommodate new knowledge, though without the kind of intellectual squabbles that were associated with specialists, and regardless of such growth in knowledge, a practical system would be judicious and consistent over time. It would not, in other words, change its notational meanings capriciously or often, forcing librarians to make similar changes in their collections. One might well sum up this point of view by saying that a practical classification was made primarily for classifying books typically found in libraries. It did not begin with an ideal development of any particular subject area, but rather with the coverage of a subject area as it might be typically represented in such a collection.

In contrast, a theoretical classification was considered to be one that began with an ideal conception of all subjects and, on the basis of that beginning point, one that would develop subject structures which were logically consistent and which were perceived to be accurate by specialists. Once such a structure had been developed, it could then be applied to various classificatory purposes, one of which was shelf classification, but another of which was a classified catalog. A classification of this kind would also pay attention to changes in subject relationships and the rise of new knowledge so that as these changes occurred and new subjects arose, they could be incorporated into the system. In short, a theoretical classification was viewed as having its basic loyalty to knowledge structures themselves rather than to library book collections. When knowledge changed, such a classification would also change, regardless of the effect that making such a change might make on how a library operated.

Considerable impetus for this public discussion came from the publication in 1905 by Paul Otlet and Henri La Fontaine of their *Manuel du Répertoire Bibliographique Universel.* The first edition of the *Manuel* would later be published as the *Classification Décimale Universelle* (CD), and eventually come to be called the Universal Decimal Classification (UDC). This system had been based on the DDC, but went well beyond it not only in schedule development but in the use of notational synthesis and collocating devices. Further, because it was developed in the context of documentation goals to assist scholars in gaining access to information in the IIB catalog, considerable effort had been taken to develop and arrange its categories in the way that scholars and specialists (especially those in science and technology) were likely to view subject arrangements. Toward this end, specialists were engaged to develop various parts of it.

Considerable impetus for the discussion was also generated by the appearance of the initial schedules of the new Library of Congress Classification (LCC). The LCC was immense in size compared to the DDC, not simply because it tended to enumerate large numbers of categories which were accommodated by synthetic devices in the DDC and in the *Manuel,* but also because it had to meet the subject arrangement needs of the nation's largest library, the collections of which already exceeded a million volumes. In a strange way, the LCC occupied both sides of the fence in the public classification discussion. On the one hand, because it had been devised specifically to accommodate books—the books of the Library of Congress—and paid only modest attention to scholarly considerations in subject collocation, it had the feel of a practical rather than a theoretical classification. On the other hand, because it was so detailed in comparison to the DDC, and because it had been structured for all practical purposes as a connected sequence of special area classifications, it struck others as more of a theoretical system.

Another factor made this overall situation even more complex. Dewey had agreed to cooperate with the makers of the *Manuel,* and had invited them as well as others to submit suggestions for revisions of the DDC. This invitation led many to suggest expansions of particular parts of the system. Some suggestions were made by librarians and arose from practical applications of the DDC. But many other suggestions came from specialists, especially in science and technology, and amounted at times to extremely detailed classification structures. By 1916 the ALA itself had gotten into the act by negotiating the creation of an Advisory Committee to the DDC. It might have been expected to champion the needs of the moderate-sized general libraries, but it was in actuality chaired for all of its existence (into the 1930s) by C. W. Andrews, chief librarian of the John Crerar Library in Chicago, and as such became the source of some of the calls for more accurate and detailed schedules in science and technology. Had the DDC adopted the changes suggested, many of its sections would have had to be overhauled completely because the suggestions often organized subjects in ways different from the DDC. This would have violated Dewey's basic tenet of the integrity of numbers, however, a move he and his editors were rarely willing to make.

The chief result of these various developments was to put pressure on Dewey, May Seymour, the editor, and Walter S. Biscoe (who continued to help on the scheme) to make extensive changes in the system. Although initial calls began to be voiced by 1916 to reduce the size of the system so as better to serve libraries of moderate size, most specific suggestions came from specialists who called for the inclusion of extensive and complex revisions. Some of the extensive revisions which made their way into the system during this period (along with their corresponding sources) are found in Table 1.[8]

Table 1. Extensive Additions to the DDC in the 7th to 11th Editions

7th Edition (1911)

540	Chemistry—John Crerar Library, Chicago; MIT, Cambridge, Mass.; and the Concilium Bibliographicus, Geneva, Switzerland.
611–612	Anatomy and Physiology (58 p.)—UDC sources.

7th and 8th Editions (1911, 1913)

620s	Engineering (various parts—more than 40 p.)—MIT, University of Illinois, Iowa School of Mines, and various practicing engineers.

10th Edition (1919)

069	Museum administration (13 p.)—Museum administrators. Required rearrangement of the 060s to make room.
630	Agriculture (47 p.)—sources not specified, but from specialists.
960s	Africa—subdivisions—An English librarian, A. J. Hawkes.

What was most impressive about each of these changes was the extent of the revisions in terms of the total numbers of categories and pages involved. They did not revoke the integrity of DDC numbers to any great extent because the changes occurred for the most part in relatively undeveloped sections. In addition to these modifications, beginning in the 7th edition the editors also included a section in the introductory pages of the system which explained the special punctuation and collocation provisions of the "Brussels expansion" of the DDC.

Other changes made in the system came from sources closer to home. These changes differed from the foregoing chiefly in extent, rarely extending beyond a half-dozen pages, the expansion in the 640s being an exception at eleven pages. The more notable ones are found in Table 2.

Ultimately, editorial work on the DDC during this sequence of editions was just able to hold the line against the pressures for change exerted on it. Even

---

[8] The sources listed after the sections come from a variety of locations in Comaromi 1976, pp. 245–96.

so, the system nearly doubled in size during this period, from 511 pages in the 6th edition in 1899 to 988 pages by the time the 11th edition was published in 1922. This growth had the effect of giving the system an uneven appearance, however. Much of the system—the 000s, 100s, 200s, 400s, 700s, and 800s—ambled along

Table 2. Other Revisions of DDC Editions 7–11

7th edition (1911)

020s	Library economy
136.7	Child study. Paidology
355–358	Military science
370–372	Education (General topics) and Elementary
640s	Domestic economy

8th and 10th Editions (1913, 1919)

320s	Political science
330s	Political economy

10th Edition (1919)

940s	Time periods for the World War
971	Canada—subdivisions

without overly extensive revision. In addition, the special tables that had first been included in the 2nd edition and revised in the 3rd were included in every edition to the 11th with barely any changes at all. In contrast to the seeming stability of the foregoing, the 300s, 500s, 600s, and 900s contained much change. Taken as a whole, these classes dramatically surged ahead of the remainder of the system in terms of number of pages and extent of detail. A side effect of this growth was to encourage still others with specialist interests to ask for revisions in their areas of interest.

Editions 12–13 (1927–1932)

The 12th and 13th editions of the DDC continued the same kind of uneven expansion that had characterized the previous five editions and did so primarily in response to the same kind of pressures that characterized the previous period. As a result the system again grew rapidly, this time by more than 650 pages. The most notable revisions of these two editions are listed in Table 3.

What made all such changes in these two editions different in character from those in the previous five editions, however, were the interrelationships that occurred in the editorial work of the system. In 1922 May Seymour died. She was replaced by Dorcas Fellows who was even more ardent in her devotion to both Dewey the man and to Dewey's classification system. In 1922 as well, the editorial office of the DDC was moved once again to Albany, New York, and in 1927 it was moved still again, this time to the Library of Congress in Washington, D.C. Dorcas

Fellows, pleased to labor for relatively little remuneration for the sheer honor of working closely with Melvil Dewey, had a very strong vision of the DDC as the best and only appropriate system designed chiefly for the typical general library. She took this vision as a faithful rendition of why Dewey had created the system in the first place. She became very protective of the system, guarding it (and, in many respects, Melvil Dewey himself) from competition (for example, in the form of the

Table 3. Notable Revisions of DDC Editions 12 and 13

12th Edition (1927)	
630s	Agriculture (an additional 22 pages)
658	Business methods. Industrial Management (from one line to 9 pages)
670s–680s	Manufactures; Mechanic trades (an additional 19 pages)
940.3–.4	The World War (an additional 23 pages of very detailed subdivisions)
Index	An additional 109 pages
13th Edition (1932)	
150s	Psychology (from a single page to 34 pages in a first schedule, with a 55-page alternative schedule at 159.9)
380s	Commerce. Communication (from 1.5 pages to more than 8 pages)
580s	Botany (from 7.5 pages to 58 pages)
952	Japan (from one line to 5.5 pages)
Table	General forms subdivisions (from 1 page to 8 pages, including extensive subdivisions for the first time, as well as material from the UDC)
Index	An additional 174 pages

LCC) and from those who in her judgment would make it serve some other purpose. In fact, she so identified herself with this vision of the DDC that she took any criticism of it, good or bad, as an affront to the purposes of the system, to Melvil Dewey, and to herself.[9]

Fellows expressed her fierce loyalty in the context of other critical changes also taking place, especially for Melvil Dewey himself. Once the 11th edition of the DDC had been published in 1922, Dewey, then entering his eighth decade, attempted to back away from close involvement with the system. In 1922 he transferred copyright and all powers over the system to the Lake Placid Club Education Foundation, and in 1923 the Foundation in turn appointed his son, Godfrey Dewey, as the administrator of the system. This was a reasonable act, of course, because someone in New York had to coordinate developments with the Foundation, and as long as it was not to be Melvil Dewey himself, Godfrey Dewey represented a strong choice. Godfrey Dewey was Harvard-educated (including A.B., A.M. and Ed.D.), understood the nature of his father's work, and had been conversant with the DDC from his earliest years. At the same time, Godfrey was clearly enthusiastic about

---

[9] Comaromi (1976, p. 279) states that Fellows tended to distrust outside groups, particularly the ALA Advisory Committee on the DDC, headed by C. W. Andrews, which she fought against "with fang and claw." In Comaromi's book and in the correspondence related to this period in the Dewey papers at Columbia University, one cannot miss the regular appearance of "embattled" statements on her part.

international cooperation in all realms, not the least of which was international cooperation in classification. His interest in cooperation focused especially on bringing together the DDC and its Brussels expansion, the UDC, created by Paul Otlet and Henri La Fontaine. Cooperation between the two was problematic, however, because the UDC was just the sort of system with which Fellows found herself uncomfortable.

The UDC had at first developed in parallel with the DDC so that there was a reasonable amount of correspondence between them. However, the correspondence began to break down by World War I and by the early 1920s the two systems already diverged considerably. When Godfrey Dewey became involved in DDC work, he and Paul Otlet suggested to Melvil Dewey that three separate editions of a single system be published—an abridged edition for very small libraries, a standard library edition for shelf classification in libraries of average size, and a well-developed "bibliographic" edition which would meet the documentation needs of specialists and be used primarily in classified catalogs. In 1924 when the IIB was reorganized, its new leaders pressed for the idea of making a new second edition of the Institut's *Manuel,* to be named the *Classification Décimale* (i.e., the UDC). Melvil Dewey agreed in principle on having three editions and in 1924 sent Godfrey Dewey and Dorcas Fellows to a conference in Geneva, Switzerland, to discuss the matter with representatives of the UDC. An agreement was made to pursue the matter further. Fellows' role in the project was to compare the two systems and harmonize them.

Fellows attempted to harmonize the schemes over the next five years but became disenchanted with the task and dragged her feet in the matter. Attempting to bring about harmony between the two systems simply demonstrated to her that the schemes had diverged too much. She also became impatient with Fritz Donker Duyvis, Samuel Bradford, and A. F. C. Pollard, those responsible for the UDC. She considered them far too theoretical to understand the practical needs of general libraries that used the system. Ultimately, she wrote directly to Melvil Dewey that she thought those responsible for the UDC were incompetent and that the plan to harmonize the two schemes was dangerous to the future of the DDC. She also came to the end of her patience with Godfrey Dewey. She identified him with the UDC leaders and refused to do many of the things that he, as the administrator of the DDC, requested of her. Matters became so strained between Fellows and Godfrey Dewey, in fact, and so frustrating to Godfrey Dewey, that Melvil Dewey had to step in more than once to calm down the situation. Finally, Godfrey Dewey resigned his position in October 1931. This left Melvil Dewey reluctantly in charge of the system once again, but his resumption of active work was ended by his own death two months later in December 1931. His death left the way clear for Fellows to pursue her own course with the DDC, and her own course included neither the UDC nor other specialist approaches to classification which would have required her to violate Melvil Dewey's original vision of the DDC.

The 12th and 13th editions of the DDC (1927 and 1932) were published in the heat of this ongoing conflict. In this respect, some of the changes already listed for these two editions in Table 3 (12th ed.—658, Business methods; 13th ed.—150s, Psychology, and 380s, Commerce) actually constituted "protest" expansions authored by Fellows in direct contradiction to other expansions suggested or offered by specialists in the fields they represented. Business and management specialists, for example, had requested a revision of 658 even before the 11th edition had appeared in 1922. But, when Fellows became the editor, she disallowed all the work already done, spurned comments from the ALA Advisory Committee, dallied with the revisions, and eventually published her own version, first as a separate publication and then in a shortened form in the 12th edition. She likewise spurned a "scientific" revision of 582 Systematic and Taxonomic Botany and published her own revision of the same section of the DDC in the 13th edition. Finally, for Psychology (150), which many concluded badly needed revision, she developed an expansion that preserved the older notations of the scheme despite advice to the contrary. Her only compromise was to include an alternative, more scientifically approved, schedule in the single notational location of 159.9. However, this alternative schedule, which reiterated the entire field of psychology in an enormous schedule of fifty-five pages, had the disadvantage of having very lengthy notations, some reaching fourteen digits.

Despite her resistance to the suggestions noted here, Dorcas Fellows did respond ultimately to the currents of the times. During her tenure, the DDC continued to be augmented with special expansions, though not perhaps with results that made specialists happy. In another less obvious way, the notational technology of the UDC began to have an effect on the DDC. This effect is evident in the special tables included in the 13th edition, especially the second table which, in this edition, was called "Common subdivisions." This table included one full page of UDC "Miscellaneous common subdivisions" and "Viewpoints," followed by two and a half pages of form divisions, now fully subdivided into a sizable number of subdivisions. In order to incorporate the UDC categories into the system, instructions were added that appear to recognize for the first time the power of the "zero" in the notation of the system.

Editions 14–15 (1942–1952)

The conflicts which underlay the 12th and 13th editions marked the end of the role played by those who had guided the system during its first sixty years. Melvil Dewey died late in 1931 and Walter S. Biscoe in 1934. Dorcas Fellows herself, always of somewhat tenuous health, became seriously ill in 1935. Her editorial work suffered, but when she died in 1938 no editorial successor had been chosen.

Beginning in 1932, other parties concerned about the future of the DDC began to assert themselves. By 1937 the advancing disability of Dorcas Fellows led to the establishment of the Decimal Classification Committee. The Committee included representatives of both the Lake Placid Club Education Foundation —in this case, Godfrey Dewey who in this way became re-involved in DDC work—and the ALA—namely, Milton Ferguson who served as the Committee's chair. Although the purpose of the Committee was to set "the management and editorial policies of the DDC" (Comaromi 1976, p. 367) over the next decade and a half, the Committee struggled to define its particular duties and responsibilities, vacillating between policy work and the more particular work of an editorial board. In 1938 the Committee appointed Constantin Mazney as the new editor of the system, but the lines of communication between Mazney, the Committee, and the DDC Section of the Library of Congress were not very clear. Lack of effective communication and the absence of a single, strong leader subsequently brought disjointedness to DDC work, and the effect of such disjointedness was to produce two editions, the 14th and 15th, which fractured its tradition of development. This was especially the case when an attempt was made during this period to produce a "standard edition."

The idea of a standard edition of the DDC had floated about in the library community for more than three decades. In the 1920s a standard edition was to have been the second of the three versions of the system envisioned by those attempting to bring the DDC and the UDC into concordance, but because the chief goal of those negotiations was to make the way clear for a "bibliographic edition," the idea of a standard edition was not defined or pursued in any systematic way. By the beginning of the 1940s, the idea of a standard edition finally began to assume some shape, but it is evident in retrospect that the idea remained very unclear and had conflicting elements.

The idea of a standard edition can be conveniently divided into those elements having to do with the system itself and those having to do with the use and users of the system. Elements having to do with the system itself focused first on its level of specification and second on its up-to-dateness. A standard edition was to be one that regularized the system's average or typical level of specification—that is, that set a standard as to how far down hierarchically the system was extended in any typical section of subjects. This required smoothing out the 'lumpiness' that the system had gained over the years, where some sections had been extended downward in many hierarchical levels (with correspondingly long notations) while other sections remained relatively undeveloped, often having only three levels of hierarchy and the minimum of three digits of notation. In order to reach this goal, the system had to be pruned in some sections and extended downward hierarchically in others in such a way that its level of specification would consistently reach some ideal level of hierarchy. The chief difficulty was to determine a desirable level, because choosing one level over others had an impact on different kinds of libraries in different ways.

The other element of the idea of a standard edition having to do with the system itself was how to keep the system up-to-date or current in both its choice of terminology and its subject collocation practices.[10] Up-to-dateness is another way

---

[10] The terminology spoken of here is the terminology used in the schedules of the system. Subject collocation pertains on the one hand to how related subjects are brought together—for example, whether one should or could bring together in one place in the system all topics having to do with TOBACCO even though under normal circumstances they might be scattered throughout the system under AGRICULTURAL CROPS, under MEDICINE, under SOCIAL PROBLEMS, etc.—and on the other hand to how the subjects in a given section are subarranged—for example, how one might subarrange all the subtopics listed under SHIPS or all the subtopics listed under the general field of ENGINEERING.

to speak of how well the system accommodates the changing and evolving subjects of the universe of knowledge. Over the years, an increasing number of terms and subject sequences of the DDC had become out-of-date, at least in the opinion of many scholars and librarians. As a result, the system acquired a reputation of being out of touch with change. This was evident especially in subject collocation matters because of the long-standing policy of maintaining "integrity of numbers," a policy begun by Melvil Dewey with the 2nd edition. To accommodate change in subject collocation, subjects had to be relocated to new numbers, which had been done sparingly over the years. The problem was how to determine the extent to which change should be accommodated, because pursuing significant revision would involve a large variety of potential impacts on the users of the system.

The users of the system could be viewed variously: 1) in terms of size of library, using a simple scale based on small, medium, or large libraries (keeping in mind that small general libraries were served by the abridged edition that Dewey had first created in 1894); 2) in terms of kinds of libraries, either general libraries which covered the universe of knowledge in their collections, or special libraries which covered one or more parts of the universe of knowledge in great depth; 3) in terms of the extent to which a library could accommodate the expense of making constant changes, using a simple scale based on no changes, some changes, or all changes; and 4) in terms of how the system was used, i.e., for shelf classification or for a classified catalog. There are two possible levels of specification: medium = 4–6 levels and their corresponding digits; and full = 7 or more levels and digits, with very brief specification addressed by the abridged edition. If the kind of changes are limited to those having to do with subject collocation,[11] the two levels of specification, when combined with the factors listed above (size, kind, expense, use), yield more than seventy different combinations representing potentially different points of view which librarians might bring to any consideration of what a standard edition should include.[12]

Given this large variety of ways to view the idea of a standard edition, it is no wonder that the idea remained confusing and vague. Nevertheless, pursuit of a standard edition controlled the development of the system during the 14th and 15th editions. This is most likely the case because Melvil Dewey himself appears to have assumed tacitly the idea of a standard edition in his business-oriented approach to the system.[13] A standard edition was one which was designed principally to serve

---

[11] Making changes in terminology is excluded because terminology changes did not really have an impact on how the system was applied.

[12] The formula for the number of potentially different points of view that librarians might bring to standard edition considerations is: 3 sizes of libraries x 2 kinds of libraries x 3 levels of accommodation to change x 2 kinds of uses x 2 levels of specification x 1 area of change = 3 x 2 x 3 x 2 x 2 x 1 = 72. For example, a large, general library that could not afford to follow any but a few changes, and that used the system for shelf classification, represented one point of view in considering the idea of a standard edition, whereas a medium-sized special library which could afford to follow all changes and which used the system for its catalog represented a far different point of view. To expand any of the scales used here with more points of difference would simply increase the number of points of view which users of the system might bring to the issue.

[13] Dewey's "business-oriented approach" to the DDC is discussed in Part II of this paper.

the shelf classification needs of the core libraries of the modern library movement in the United States, namely, public libraries of moderate size. His sense in the matter was understandable, of course, because these libraries represented the predominant market for the system. Later, as these libraries grew in average size over the years, Dewey expanded the system enough to keep up with their needs. He did so, one imagines, in order to retain his dominance over the market. When the needs of these libraries outstripped the needs of newer or much smaller general libraries, he created an abridged edition of the system for the latter.

Dewey also approved other expansions of the system, but these appear to have been made chiefly to placate users of the system who did not find the standard edition adequate. Dewey never seems to have been very comfortable with those kinds of changes. What he seemed unable to envision was one or more alternative editions of the system for libraries which were much larger than the average library, or which were specialist rather than general in orientation. Neither was he able to envision a single maximum edition which could be adapted for any kind of library merely by using less or more of the system as needed.

When planning for the 14th edition got under way seriously in 1938, the immediate goal was to have it serve as a precursor to a standard edition which would be created in the 15th edition to come later. However, since the idea of a standard edition remained vague, Mazney pursued his own sense of the idea, and for him this simply meant evening up the level of specification in the system. To begin with he pruned the system to a certain extent, and thus made the first significant cuts in the history of the DDC. In doing so he reduced the system by more than 70 pages of schedules. For example, Museum administration (069), [Religious] Associations (267), and a long technical expansion of Textiles (677) were greatly reduced; and the alternative Psychology schedule at 159.9 was abandoned altogether, as were the specifically UDC parts of the "Common subdivisions" table.

This was as far as reduction of the system went, however. Elsewhere, his reductions were more than offset by additions he made to the Index (97 pages) and by other expansions. Expansion was especially evident in Fine arts (700–779) where an original 13 pages were expanded to 147. In addition, the 35-page specialist expansion of Systematic and taxonomic botany (582) which had been spurned by Dorcas Fellows for the 13th edition was included at the end of the volume as a special schedule. Although the level of specification in the system became more consistent as a result of these expansions, the overall system grew to a record-breaking 1,927 pages. These expansions troubled those for whom a standard edition meant serving the needs of medium-sized general libraries with an edition of modest size.

In pursuing the foregoing path, it seems obvious that Mazney adopted a very extended level of specification in the system and that he viewed a standard edition in turn as something appropriate for very large libraries. But, the DDC Committee apparently desired the opposite—a lower level of specification for libraries of lesser size. Consequently, the Committee was unhappy with the result and dismissed Mazney. They did not appoint a successor immediately, however, and continued lack of coordination caused work on the system to languish and remain disjointed. Eventually, Milton Ferguson, who had chaired the Committee since its inception in 1937, became the editor in 1949. He seems to have been the one most responsible for instituting a very different idea of the goal to be attained. The result was published in 1951 as the 15th or "Standard edition."

The 15th edition virtually shocked the library community because of the drastic nature of the changes it represented. The first major change was to limit subject specification in the system to between four and six levels of hierarchy. This meant limiting class numbers to six digits (with a very few slipping through at seven), although many were only three to five digits in length. The effect of this change alone was to radically reduce the entire system in size. One way to measure the change is in page counts. The schedules were reduced from approximately 1,050 pages in the 14th edition to only 467 pages in the 15th, and the index from more than 750 pages in the 14th edition, to only 181 pages in the 15th. The radical nature of the reductions can be seen especially in Medicine (610), reduced from 84 to 10 pages, in Engineering (620), reduced from 59 to 29 pages, and United States History (973–999), reduced from 98 to 7 1/2 pages.

In reality, the reductions were even more severe than these page counts suggest. The 14th edition, like other editions before it, had packed its pages by using a relatively small typeface. In contrast, the 15th edition was set in much larger type with large margins and much open space. The resulting reduction of the number of unique categories listed was very severe. Enumerated class notations had been reduced from more than 31,000 in the 14th edition to about 4,600 in the 15th edition, the size, in effect of the 7th edition in 1911 or even the 6th edition in 1899. Moreover, many third hierarchical level numbers had been omitted (274 by one count), but some of their subdivisions had been retained (Eaton 1952, p. 747). The latter led to such notational discontinuities in expressiveness as Invertebrates (592) having as its first subdivision the category Protozoa at 593.1, and the schedule moving from 985.4, the last element of Peru, to 986.1, the number for Colombia.[14] Gone as well were many of the special features of previous editions—for example, names of authors enumerated under literary time periods and genres, names of lower level political subdivisions of individual states of the United States and many other countries, and all of the special tables except Form divisions. Even that table had been reduced from its more than five pages with enumerated subdivisions to a single page listing nine numbers (01 to 09), each with only a modest number of unenumerated but associated terms (i.e., subterms) merely included in explanatory paragraphs.

Reducing the size of the system in this way and evening up the distribution of categories to rarely more than six digits of hierarchical levels fulfilled the first of the two elements of a standard edition which had to do with the system

---

[14] Expressiveness is the capacity of a system of notational symbols to portray the relationships between the verbal categories in a library classification system. Thus, if the categories list a subordinate class after a superordinate class, an expressive notation would be one that "expressed" by its symbols that relationship. A decimal fraction such as is found in the DDC is ordinarily considered to be expressive of hierarchical relationships of that kind to an infinite extent, though it is expressive of coordinate relationships only for nine subclasses, the number of direct subdivisions of any given number.

itself. But the 15th edition also attempted to deal with the second basic element of a standard edition as well—its modernization. The editorial staff attempted to modernize the system in two distinct ways. First, terminology was modernized not only by ridding the system of Melvil Dewey's reformed spelling (also championed by Godfrey Dewey), but also by replacing all old-fashioned terms with up-to-date terminology. Second, many relocations were incorporated into the system, the only indication of these being the phrase, often in very large typeface, "It is recommended that [a given category] be classed in [a different location]" placed at the original location of the category. In short, the system violated the long-standing practice of adhering to integrity of numbers by changing the collocation of many subjects.

The edition elicited some praise, especially for its modernization. But such praise was generally drowned out by biting criticism. The criticism arose from both practical and theoretical perspectives. In the practical realm, the edition was found to be unsuitable for large libraries because it was so greatly reduced in overall size. Nor was it found to be suitable even for its target audience of medium-sized libraries because of its relocations. In the theoretical realm, the chief criticism was that the manner in which the system had been reduced had destroyed its cohesion as an integrated system of categories of knowledge combined with an expressive notation. One critic, Thelma Eaton, concluded that "the classification was not considered as an outline of the terms relating to subjects but merely as a list of numbers which could be used as shortened symbols for books," thus fulfilling the statement in its introduction that the system "aims to provide a system of classifying books, not a detailed outline of knowledge" (Eaton 1952, p. 747). In Eaton's view, the changes made it impossible to teach aspects of classification theory from the system, one of the chief strengths of earlier editions.

Criticism of the 15th edition was so intense that it was reissued with revisions early in 1953 (with a 1952 copyright date). Revisions included more than doubling the size of the index and modifying some of the more radical class number relocations. However, by the time this had been accomplished the damage had been done. Over the next decade and a half a flight from the DDC system occurred, especially among academic libraries which reclassified their collections to the Library of Congress Classification system. Edith Scott later suggested that the 15th edition came close to being a "death blow" to the system (Scott 1960, p. 16). Comaromi concluded, however, that the 15th edition had a positive as well as a negative side. Having severed the system from its past, and having modernized its terminology, collocation of subjects, and layout, the 15th edition made it possible to turn the system in a new direction.

Recovery and Advance: Editions 16–21 (1958–1996)

The most recent period in the development of the DDC, one with which most of us are familiar, has seen the recovery of the DDC from the worst effects of the 15th edition and its advance into realms well beyond anything previously envisioned. The 16th edition, published in 1958 and edited first by David Haykin and later by Benjamin A. Custer, recouped the losses of the 15th by restoring the system to something of its previous size. But it also retained the modernization begun in the 15th edition. In editions since the 16th, the system has undergone massive growth. The evolution of the system into its present form is due in great part to the work and ideas of Benjamin A. Custer. In addition to his work on the 16th (1958), he was editor of the 17th (1965), 18th (1972), and 19th (1979) editions. There were further developments under the leadership of John Comaromi in the 20th edition (1989) and under the leadership of both John Comaromi and Joan Mitchell in the 21st edition (1996). Because recent developments in the DDC are much more generally familiar to us, no edition-by-edition accounting of this period will be necessary. However, two significant aspects of DDC development in this period do merit special comment—the accommodation of classificatory change in the DDC and the incorporation of modern library classification techniques.

Accommodating Classificatory Change

Accommodating classificatory change refers to keeping a classificatory system like the DDC up-to-date in its terminology and, especially, in its collocation of subjects. Both of these aspects arise in turn from the dynamic and changing nature of the universe of knowledge itself. New knowledge arises continuously and older knowledge often shifts in its relationships with surrounding subjects. Thus, any scheme which purports to represent the universe of knowledge and its relationships must accommodate this dynamic and changing scene. The failure of earlier editors of the DDC to pursue classificatory change in a systematic and rigorous way underlay a great deal of the conflict of the second period of DDC development. In the present period, the situation has been reversed. Classificatory change has been pursued vigorously in the DDC, although not without also attempting to soften its most troublesome effects on libraries which use the system.

Accommodating classificatory change at its base means incorporating in the system those new subjects and subject subdivisions which match the growth of knowledge as represented in library collections and which are of interest to library users. During this period, this kind of accommodation is evident in the DDC by the sheer increase in the size of the system. The 16th edition restored much of the categorical structure of the system which the 15th edition had omitted. It ended up with nearly 18,000 entries, in comparison to the mere 4,600 entries of the 15th edition, though this total was not as large as the more than 31,000 entries of the 14th edition.

In this kind of count an entry refers to a uniquely enumerated class number, and the total number of entries in a particular edition refers to all of the uniquely enumerated class numbers in that edition. In the past, counting entries in the system had been a useful way to compare the size of editions. Beginning with the 17th edition, however, a count of the total number of enumerated entries in the system could no longer be used as an indicator of the actual size of the system. This is because of the growing use of synthesis in the system. Synthesis allows parts of class numbers or numbers for categories in special tables to be combined with enumerated class numbers in the schedules in order to create still more class numbers, thus leaving much of the system unenumerated. Thus, a count of only the enumerated entries in the system does not reveal how large the system really is. That the system has grown is evident, of course. As to its ultimate size, it is sufficient to say that synthesis makes the present size of the system exponentially larger than any edition in the past.[15]

Accommodating classificatory change, besides involving the addition of new subjects and subject subdivisions, also means relocating subjects in the system when that is appropriate. The need to relocate a class number arises when that class changes its conceptual alliances over time. Relocation can be accomplished either by changing a topic's class number completely or simply by shortening or lengthening its class number. The editors of the 15th edition made many relocations in order to catch up with changes not previously pursued. In the process, some of the changes they made were later considered to be excessive or unwise. The editors of the 16th edition found it necessary to undo some of the relocations made in the 15th edition as well as to make still others. Consequently, the total of approximately 1,600 changes in the 16th edition was quite high. Since then, the total number of such changes has dropped with each edition. The 17th edition had nearly 750 changes, the 18th nearly 650, the 19th some 400, and the 20th only somewhat over 300. Since the 18th edition, these numbers have also included changes in special tables as well as in the schedules.

Some changes are more significant than others. Beginning in the 16th edition, changes have been incorporated in certain sections of the system (not included in the above counts) by significantly redesigning their entire structures. Revisions of this kind, sometimes called "phoenix" schedules after the legend of the phoenix bird, are listed in Table 4 for Editions 16 through 20. In the 21st edition, thoroughgoing revisions of this kind have been even more extensive. Public administration (350-354) and parts of Life sciences (560-590) have been completely revised, while Education (370) has been extensively revised.

---

[15] A simple calculation is sufficient to demonstrate that the system has grown exponentially in size. The 19th edition, the last to provide figures for the number of enumerated entries, listed 21,504 entries in the schedules (not counting instructions for expansion by some synthetic device) and also separate counts of each of the entries in the seven special tables of the system. Since any number can potentially have the 116 listed categories of Table 1, Standard Subdivisions, attached to it in order to create further subcategories, and since Tables 5 (192 entries), 7 (440 entries), and 2 (6,933 entries x 5 + ca. 6,500 x 2, because the entire Table 2 can be added to T1 in at least four places in —07 and, of course, in —09 as well, and most of Table 2, —3-9, can be added to two different places in —07) can subsequently be added to Table 1 at specific locations, then any count of potential classes in the system must also take into account these potential synthetic additions. A formula for calculating even this minimal approach to potential classes in the system, would be 21,504 x [116 + (5 x 6,933) + (2 x 6,500) + 192 + 440] or 21,504 x 48,413 = 1,041,073,100. Since 31,000 is the largest number of entries ever enumerated in the system (14th edition) and $31,000^2$ is 961,000,000, it is evident that even when only a minimum number of synthetic devices are provided, the size of the system has already increased exponentially by more than the simple square of its greatest total of enumerated unique class numbers.

Other less obvious changes have included shifting the location of certain aspects of subjects. For example, in the 19th edition, modern political history, formerly at 320.9, as well as the civilization and culture of a specific place, commonly classified in 914–919, were formally incorporated together into the history of a specific places at 940–990. And now in the 21st edition, the first significant attempts to redress long-noted occidental biases in the DDC are the basis of significant revisions of the 200s (Religion). Finally, keeping up with change is also evident in a continuing quest to use appropriate terminology in both the schedules and index.

Table 4. Phoenix Schedules in the 16th through 20th Editions of the DDC

16th edition
546–547   Inorganic and Organic chemistry

17th edition
150   Psychology [Since many of the categories of the revised 150 schedule were taken from the 130s, what remained of the 130s also amounted to a revised schedule, though mostly because of omissions.]

18th edition
340   Law
510   Mathematics

19th edition
301–307   Sociology
324   The Political Process [The new 324 schedule was a combination of the old 324 and 329 schedules. Subsequently, the 329 section has remained emptied of content.]
41–42   United Kingdom area table numbers

20th edition
004–006   Data processing and Computer science
780   Music
Table 2   Many locations

The pursuit of classificatory change of the kind involving relocations, when viewed on an edition-by-edition basis, has actually been pursued with a great deal of constraint. Constraint has been exercised in part because of the potential disruption relocations have for the libraries which use the system, especially for shelf arrangement. Relocations require changing catalog entries as well as the class numbers put on physical items such as books, sound recordings, maps, etc. Where the number of such changes is large and when many such changes come all at once, the resources necessary in an individual library for making the changes may be inadequate. Constraint has also been exercised because some changes need more compelling evidence before being adopted than might at first seem obvious and this has the effect of delaying decisions about them. This is done to guard against hasty decisions about changes which might eventually have to be remade. In the worst cases, decisions of the latter kind have led to the specter of "wandering subjects." Wandering subjects are subjects which have moved from one place to another, and sometimes back and forth between the same two places, over more than one edition of the system.[16]

One obvious aspect of accommodating classificatory change is the amount of time and effort—and, therefore, cost—that libraries are required to expend in implementing the changes. This expenditure is evident, on the one hand, in the need to make changes in the class numbers of existing works in a library's collection and, on the other hand, in the need to learn how to work with a scheme which has become more complex with each edition.

Three counterbalancing factors have somewhat lessened the impact of such change. One such factor has been the publication of changes as they occur rather than allowing them to accumulate for publication only with the advent of each new edition. Advance publication of changes has been made possible both in the form of *DC&*, a publication begun expressly for this purpose with the 16th edition, and in the form of separately published sections of the system (for example, 004–006 in 1985). Timely publication of changes has the potential of spreading the activity of reclassification over a longer period and therefore of more readily integrating it into a library's ongoing technical services.

Another factor which has lessened the impact of change has been the increasing availability of electronic catalogs where making changes, at least in the catalog records of items, is less cumbersome than having to make the same changes on paper records. Further, to the extent that the same automated systems can produce labels and other materials helpful in the physical processing of materials, the same systems also increase the ability to handle changes for the physical materials themselves.

A third factor which has lessened the impact of change has been the increasing availability of "classification copy" for library materials, not only that which originates at the Library of Congress but also that which arises from cooperative ventures such as OCLC's Online Union Catalog.[17] As copy has increased in amount, the necessity for librarians to provide original classification numbers for the same works locally has decreased. The value of classification copy is particularly evident for classifying materials under a new DDC edition which has incorporated more complex classificatory operations than those found in a previous edition in given sections of the scheme—for example, by incorporating faceting formulas where formerly there were none. In short, it circumvents some of the need to contend

---

[16] The migrations of various subjects are pointed out by reviewers from time to time —for example, the movement of Cultural anthropology from 572 (up to the 16th edition), to the 390s (17th edition), and finally to 301.2 (18th edition and since); the topic Prehistoric archaeology from 571 (up to the 16th edition), to 913.03 (17th edition) and finally to 930.1 (19th edition and since), and the topic Criminal offenders from 364.38 (14th edition), to 364.13–.17 (16th edition), to 364.3 (17th edition), and back to 364.13–.18 (18th edition). But, all things being equal, instances of this kind of wandering are few.

[17] "Classification copy" refers to classification numbers for individual information-bearing entities which have been supplied by classifiers other than the one who is looking for such numbers. Here it specifically refers to DDC numbers supplied for items. The phrase is simply an extension of the term "cataloging copy" which refers to cataloging data for particular items supplied from some source other than a particular cataloger.

with the growing complexity of the system. One may also include at this point the help afforded by the Library of Congress Decimal Classification Division program of providing segmented classification copy. Segmented class numbers provide a basis for reducing the length of classification numbers in a logical way for their use in a local collection. Providing a ready-made template for making such reductions in length simply speeds up the process of using classification copy.

Alleviating the impact of accommodating classificatory change for the libraries which use the DDC presupposes, of course, that librarians want to keep up with such changes. This is not always the case, however. In the past some libraries would persist in using older editions in order to avoid making changes associated with a new edition or even to avoid purchasing a new edition. The extent to which this practice is followed is not known. But where it is followed, it hinders the use of classification copy which tends to be based on the most recent edition available.

Other libraries, having begun to use a newer edition, often make few of the changes required by the edition regardless of how this might affect the use of the system in their libraries. The usual reason given for not making changes is that the library simply does not have the financial or human resources to do so. This does not seem to be a fully viable excuse, however, especially given the factors that have made it easier to accommodate changes. An investigation of this phenomenon would be merited, especially one that examines how classification schemes are used in libraries and analyzes attitudes towards the growing complexity of DDC classificatory structures.

Incorporating Modern Library Classification Techniques

The second major feature of the most recent period in DDC development has been incorporating modern library classification techniques into the system. Classification techniques refer here primarily to those having to do with synthesis, though when viewed broadly they also include other matters, not the least of which is converting library classification systems to computerized form.

Synthesis in a library classification involves merging the notations (and, accordingly, the meanings of the notations) of two or more separate lists of categories. One of those lists is commonly a set of categories enumerated in the basic schedules of the system. The others consist of any number of special tables of the values which make up aspects (i.e., facets) of information-bearing entities. The latter typically reflect attributes of the content of information-bearing entities or attributes of the containers of such contents. The second kind of table is normally devised by accumulating through division or some other method as many values for the particular facet as are possible or appropriate. For example, the area table in the DDC represents the facet or aspect of subjects called "place" and is ordinarily used to indicate the spatial limits of a topic as treated in an item. The entire list of places consists of a family of terms which are actually divisions of terrestrial and extraterrestrial places in terms of their geopolitical boundaries. It is highly detailed for terrestrial places, of course, and hardly detailed at all for extraterrestrial, the latter not going beyond the solar system of which Earth is a part. Other such lists are actually imbedded in the main schedules of the system. For example, a list of metals found in 546 Inorganic chemistry may find use in other parts of the system where metals form part of the aspect of a topic. Using the list is a simple matter of combining the variable notations which indicate the metals within the list (i.e., not including the part of their notation which signifies Inorganic chemistry) with notations elsewhere in the system.

The DDC had adopted some synthetic devices from the start, especially by its 3rd edition in 1888. These included repeating patterns of form divisions, genre divisions in literature, form and topical subdivisions in language study, geographical division, and even global division (i.e., dividing a single category by any other category in the system). However, despite the "synthetic" nature of these various devices, they were not primarily viewed in terms of the "additive" nature of synthesis but rather as mnemonic aids. Their use was slowly extended over the years, but they were improved upon in only minimal ways from the 3rd through the 15th editions. For example, in the 14th edition one will find a number of instructions to divide one category by the subdivisions of another (often close by in the schedules), but this was not a consistent or widely used practice. The primary tendency was to enumerate lists of subdivisions whenever they were needed.

The only device that seems to have been given further thought was the list of form subdivisions, expanded in the 13th edition to a complex list of subtopics, with some attention paid to the role of the zero in their notational synthesis. Also, the —09 subdivision was formally listed for the first time in the 13th edition as capable of being subdivided by place. This capability had been implied in earlier editions, but not specifically stated. Finally, the form divisions were augmented by two additional tables of UDC point-of-view categories. But the editors appear to have included these tables only with misgivings. In fact, the editors removed the UDC categories in the 14th edition, and the application of all dividing devices in the 15th edition, including the form subdivisions, was greatly curtailed.

In the present period of DDC development, synthetic devices have not only reappeared but have been extended to all parts of the system in such profusion that one can only use the system with facility by learning how such devices function. The 16th edition of the DDC (1958) restored the synthetic devices common to previous editions of the DDC but, following Melvil Dewey's earlier introduction to the system, referred to them merely as mnemonic aspects of the system. Furthermore, the use of such devices was limited in this edition. For example, the introduction to the 16th edition gave general directions for the use of form subdivisions throughout the system, but cautioned against their wide use. Further, reference to such subdivisions was limited within the schedules themselves. Other forms of synthesis—for example, for the subdivisions of language study, of literature, and by place, were simply referred to in the system as a matter of "dividing" one subject by the subdivisions of another. This practice had always been true of geographical subdivisions, of course (i.e., divide the subject at hand by the subdivisions of history in 940–999), but here, in contrast to earlier editions, no attempt was made to provide tables for some of them. The only kind of device that seems to have grown in use in this edition was that which occurred when the classifier was instructed to divide one subject by the subdivisions of another. But, even here its use was limited.

The 17th edition (1965) instituted three significant changes. First, this edition not only separated areas from the history schedules and placed them in a second table of general application, but also described the process as one of "adding" rather than "dividing." In doing so, the process was formally described for the first time in terms of its additive or synthetic nature. Second, the introduction to the 17th edition formally related "Division by more than one principle" to the word "facet" and spoke of the need to provide an order of precedence when more than one facet of a subject was evident in an item being classified (p. 45).

Third, the 17th edition also addressed the need for a citation order in places where more than one facet was involved. For the standard subdivisions this involved broaching the issue of how many zeros were needed so that standard subdivision materials were always placed in a sequence that fell immediately after the main number of the class and prior to any other divisions. In other places, it involved the use of a centered heading under which multiple "divide like" instructions were given for an entire range of categories. For example, at 633–635 Production of specific crops, instructions were given to divide the topics affected first by crop production characteristics (from 631.5) and secondly by plant injuries, diseases, pests, and their control (from 632). The latter was even marked with a facet indicator, in this case the digit "9." For example, the number for corn, 633.15, could be expanded by the subdivision "78" from 632 (representing boring insects), but only with the interposition of the "9," the resulting number being 633.15´9´78. In most other aspects of faceting and synthesis, however, the 17th edition referred to the process as one of division, not one of addition, and limited the process to a single kind of characteristic.

The 18th edition of the DDC (1972) extended the idea of synthesis enormously by creating all seven of the special tables that are now a familiar part of the system and using them formally as elements of "add" instructions in the schedules. One also finds the first use of the standard subdivision notation —04 General special for building a facet within the context of that table rather than assigning it to a "zero" subdivision in the regular schedule.[18] In addition, one finds the first instances of "in-schedule" tables—that is, tables of categories in the schedules under centered headings which were to be added to a limited range of notations in the system—at such locations as 222–224 Specific parts of the Old Testament, 329.1–329.8 Specific political parties of United States, 616.1–616.9 Specific diseases, etc.

The 19th edition (19791) continued the trend toward synthesis by explaining "facet" formally for the first time, by extending the number and size of "in-schedule" tables of limited application, and by expanding even more forcefully the power of Table 1 Standard Subdivisions. The category "collections," assigned to

---

[18] The —04 standard subdivision had been emptied of its contents in the 17th edition.

standard subdivision —08 since the 2nd edition, was reassigned to the main or "base" number of a topic. And the standard subdivision —08, now free for some other use, was reassigned to the category, "History and description of the subject among groups of persons." Thereafter, provisions were made for attaching Table 7 Persons, and Table 5, Racial, Ethnic, and National Groups to standard subdivision —08 at —088 and —089, respectively. Advances of this kind did not often allow one to indicate multiple aspects of a topic, however. Thus, one had to choose one such aspect from among different aspects of a subject which characterized it. The classifier was helped to make that choice by means of "tables of precedence" (called "tables of preference" in the 21st edition) placed in strategic places in the schedule and at the beginning of Table 1. Finally, for the first time as well, a formal faceting structure was outlined, in this case in the "Options" included in the 340s schedule for Law.

With the 20th edition in 1989 and the 21st edition in 1996, facet structures which allow one to combine more than one aspect of a subject have become a more regular occurrence. The 20th edition incorporated a structure of this kind in 780 (Music), where both the "0" and a "1" are used as facet indicators. The 21st edition has adopted faceted structures for two of its three major revisions, 350–354 Public administration and 560–590 Life sciences. And both editions have extended the power of Table 1 Standard subdivisions by allowing selected categories from the same table to be added more than once in a single class number.

As a by-product of the adoption of classification techniques of the kind described here, the DDC as a system has become increasingly more complex to use. One cannot avoid such complexity if a scheme is to include synthesis, which means that a person who wishes to use such a scheme must have some basic knowledge of the devices and how they work. At the same time, the editors of the system have also attempted to add helpful devices to the system that offset this growth in complexity. Benjamin A. Custer addressed this issue by rewriting the introduction to the system in the 17th edition in order to explain its growing technical and theoretical basis, and thereafter by rewriting sections of the introduction to provide users with better explanations than he was at first able to supply. A Manual, published separately, was written for the 19th edition, but since the 20th edition the Manual has become an integral part of the system. From the start, the Manual has included not only normal textual discussions of choices among classificatory locations, but also explanations (with flow charts) of faceting procedures. In addition, a machine-readable version of the system was created for the 20th edition, and with the 21st edition this tool includes a Microsoft Windows™ environment. All such devices have tended to lessen the impact of the DDC's growing complexity and will likely prove to be the pathway to even further developments.

Summary

This survey of the development of the DDC has traced its growth from its unique and striking formative period between the 1870s and the end of the nineteenth century, through a period of growing conflict which reached a climax in the publication of its 15th edition in 1951. From that point to the present, the DDC went through a period of recovery and advance made notable by its accommodation of classificatory change and its incorportion of many modern library classification techniques.

One obvious factor in this development has been the relationship of the DDC to the idea of the universe of knowledge. The idea of the universe of knowledge was present when the DDC was born, played a role during the conflicts of the DDC's middle period, and has obviously become very important for the expansion and growth of the DDC in its most recent period. Of course, merely pointing out that the DDC has maintained a relationship to the idea of the universe of knowledge does not explain in any broader sense the nature of the relationship or why the relationship has taken the path it has. And it certainly has not dealt with our original question as to why Melvil Dewey created this approach to bibliographic control and not some other approach. To understand these matters, we must look at a range of issues that are broader than the system itself. This will be the content of the remainder of this discussion.

PART II

The Relationship of the DDC to the Movement to Classify Knowledge and the Sciences

The classification of information-bearing entities is as old as libraries themselves. Evidence of attempts to group such artifacts can be found among collections of clay tablets in ancient times. These older efforts to classify information-bearing entities were characterized by their relative simplicity, where categories commonly reflected practical storage expediencies such as the size of the items or contemporaneous educational curricula. It was not until the post-Renaissance modern period, and especially the late nineteenth century, that library classification achieved anything of the complexity that is now associated with it, especially in theory and techniques. The origin of complexity in present-day library classification (including the complexity of the DDC) is ordinarily viewed as the result of a firm connection between library classification and a much broader movement in the modern period to classify knowledge and the sciences. In short, library classification is viewed as having grown out of that larger movement in something of a seamless development.

I believe this conclusion is questionable for library classification in general and for the DDC in particular. The purpose of this second section is to show why. First, I shall describe the earlier movement to classify knowledge and the sciences. Next, I will examine the evidence typically used to establish a firm connection between library classification and that broader movement, including claims ordinarily made to link the DDC to that larger movement. I will then compare the goals of library classification and the goals of the movement to classify knowledge and the sciences. Finally, I will describe the specific point of view that Melvil Dewey had towards the DDC—his sense of business and practicality.

The Movement to Classify Knowledge and the Sciences

A movement to classify the universe of knowledge in a new way first arose in the seventeenth century among persons like Tommaso Campanella (1568–1639) and Francis Bacon (1561–1626). These men were among the first to recognize the passing of a medieval way of thinking about the nature of knowledge in society. Because of the growth of new knowledge, they found it necessary to reconceptualize how that knowledge was structured. Their task was subsequently advanced by the encyclopedist movement of the seventeenth and eighteenth centuries. The encyclopedists attempted to throw a conceptual net around the universe of knowledge by providing summaries of its essential elements. The hierarchically structured chart of those elements created by D'Alembert for Diderot's *Encyclopédie* was one of the first classificatory charts of knowledge made. These efforts to portray what the universe of knowledge included, or to describe what its elements

33

actually dealt with, have been described by Jose Ortega Y Gasset as the rise of humankind's knowledge based on human intellectual authority in contrast to knowledge based on the authority of law or faith.[1]

An important shift in terminology should also be noted. During the seventeenth and eighteenth centuries the classificationists of knowledge had not yet settled upon the word "science" to differentiate knowledge that had become "established" in society through rigorous intellectual methods from knowledge based on opinion or faith alone. By the nineteenth century, however, the sense of a special kind of knowledge, of social knowledge that had gone through the winnowing process of establishment and approval, came to be labeled scientific knowledge as opposed to, say, mere opinion or faith. Thus, in the nineteenth century those engaging in the ordering of knowledge, came to characterize their work as the classification of the sciences rather than simply the classification of knowledge.[2]

During the nineteenth century, the classification of the sciences became an activity of enormous proportions among a wide number of participants. I sometimes speak of it as a time when anyone who was anybody in the realm of scholarship wrote a treatise on the topic. Those who participated included physicists and other scientists such as André-Marie Ampère, Neil Arnott, Wilhelm Wundt, and Karl Pearson, and a variety of philosophers of all kind such as C.-H. Saint-Simon,

---

[1] Although Campanella is best known for his *City of the Sun* (published in 1623), a utopia of the kind associated with Sir Thomas More, his ideas about the classification of knowledge are found throughout his philosophical works such as *Philosophia Sensibus Demonstrata* (published in 1591) and *Universalis Philosophiae, seu Metaphysicarum Rerum Iuxta Propria Dogmata , Partes Tres* (published in 1638). His approach to knowledge categories is summarized in Flint (1904), pp. 99-103. Bacon's approach to knowledge categories is found in his *Advancement of Learning* (first published in 1605, but republished in a revised and enlarged Latin version in 1623). Flint discusses various writers but sometimes with a maddening absence of the particular sources he used. A relatively brief discussion of the movement to classify the areas of knowledge can be found in Machlup (1982), No writers during this early modern period made visually useful charts of their categories of knowledge but rather discussed such categories discursively. The first such chart did not appear until the one found in Jean Le Rond d'Alembert's "Discours préliminaire des editeurs" in the first volume of the *Encyclopédie, ou Dictionnaire raissoné des sciences, des arts et des métiers, par une société de gens de lettres. Mis en ordre et publié par M. Diderot, . . . et quant à la partie mathématique, par M. d'Alembert* (Volume 1 published in 1751, but the entire work not completed until 1769). For the "discourse," see D'Alembert (1963). In a provocative analysis of how librarians came to acquire a public mission, Ortega y Gasset (1975, pp. 197-200) suggests that what makes the idea of knowledge in the modern period different than that of preceding centuries is that for the first time it focused primarily on knowledge of human discovery and authority rather than knowledge that gained its authority from the state (i.e., knowledge based on law) or from the Church (i.e., knowledge based on theology and the dictums of a religious establishment). He signified the three kinds of knowledge sources and authority by the metaphor the book as a book by an author rather than the book as a revelation or a code.

[2] Even this sense of science as established knowledge was much broader than the idea of science in the twentieth century, however. In the nineteenth century, science still encompassed a very broad range of intellectual endeavors. It tended to include any knowledge established through rigorous intellectual or experimental method. This broad sense of the term changed during the twentieth century, however, as a result of the rise of such currents of thought as logical positivism, the development of behaviorist methodology in the human sciences, and a steady growth in emphasis on quantification in research. As a result, science in the twentieth century has come to be more narrowly viewed as endeavors whose methodologies are empirical, experimental, and quantified. This has led in turn, at least in the United States, to distinguishing scientific knowledge per se from the knowledge produced by the social sciences and by other realms of study.

Auguste Comte, Herbert Spencer, William Whewell, Tomas G. Masaryk, and Friedrich Engels. The list could easily be extended.[3]

In his examination of the movement, R. G. A. Dolby provides five reasons why interest in the classification of the sciences was so intense during the nineteenth century: 1) the success of classification schemes in botany and zoology in the eighteenth and early nineteenth centuries; 2) the sheer number of classification schemes available which caused many to take notice of the matter; 3) the rise of encyclopedism in the preceding period; 4) the general concern of scientists to characterize their fields; and 5) the rise of an inductive philosophy of science in which classification as a method played an important role (Dolby 1979, pp. 171–74). The resulting knowledge classification schemes by virtually dozens of writers ran the gamut of philosophical points of view and results.

Library Classification and the Classification of Knowledge and the Sciences

In the twentieth century, the library classification field has tended to assume that there is a firm connection between library classification and the movement to classify knowledge and the sciences. The acceptance of this connection is evident in the extended analyses of the nature of library classification by E. C. Richardson and Henry E. Bliss earlier in this century. It is also evident in the rise of the documentation movement and seems to be assumed in the thoughts of people like Samuel Bradford and A. F. C. Pollard, both of whom placed scientific classification at the apex of library work. More recently the work of some in both the British Classification Research Group and in the German knowledge organization community have occasionally tended to support a strong connection between the two realms.[4] Finally, the idea of a firm connection also echoes strongly in the pages of E. I. Shamurin's extensive history of library-bibliographic classification, although in his view a truly scientific classification must be built on a foundation of dialectical materialism.

I myself have long accepted the idea of a firm connection and over the years have made it a point to state that modern library classification arose from the "seedbed" of the movement to classify knowledge and the sciences that existed in the seventeenth to nineteenth centuries (Miksa 1994, p. 147).

On the face of it, there seem to be good reasons for assuming this connection. To begin with, both those classifying knowledge and the sciences on the one hand and library classificationists on the other hand were dealing with categories of knowledge. In fact, when one compares library classification with knowledge classification, library classification systems of the time actually outshone, in their detail

---

[3] Listings with discussions can be found in the work of E. C. Richardson (1930), Robert Flint (1904), and E. I. Shamurin (1955–59).

[4] B. C. Vickery (1975), a principal member of the Classification Research Group in Great Britain, provides a lengthy discussion of nineteenth-century classifications of the natural sciences, but connects them to the twentieth century only in the work of E. C. Richardson and H. E. Bliss.

and complexity, the systems created by knowledge classificationists. Even the 1st edition of the DDC, for example, was developed well beyond the typical philosophical scheme. Furthermore, it was no secret that countless classified catalogs, at least in the United States, had been patterned after the work of Francis Bacon, either alone or as mediated during the eighteenth century by D'Alembert in the *Encyclopédie*.[5]

In addition to such general comparisons, some library classificationists made explicit note of the connection. Charles A. Cutter, for example, lauded by his contemporaries for the scientific accuracy of his classificatory work and well acquainted with John Fiske, the American populizer of Herbert Spencer's philosophy of evolution, was quite straightforward in asserting in 1897 that his Expansive Classification was based on the idea of evolution, a basic theme of Spencer's philosophy (Cutter 1977a). In an 1880 article, Cutter had also discussed somewhat philosophically the arrangement of the natural sciences in his library classification scheme. In doing so he gave credit for some of his ideas to the scientist Richard Bliss, who was the curator of Harvard's ichthyology museum, and who regularly helped Cutter with his classification work (Cutter 1977b).

Despite such evidence, I have more recently come to the conclusion that the assertion of a firm connection between the two realms has serious problems. The issue is not whether there was any connection at all. No one can easily doubt that. The issue has to do with the nature of the connection. I have concluded that the connection is actually very weak, and the weakness of the connection has serious ramifications with regard to the way in which we view the growth of library classification since the late nineteenth century.

The beginning point for my conclusion is very simple. No matter how hard we might try to see it, there is no essential connection between the work of the original library classificationists of the modern era—Melvil Dewey, the personnel at the Library of Congress, and Charles A. Cutter—and the movement to classify knowledge and the sciences. By an essential connection I mean one in which library classificationists did more than simply adopt this or that set of categories from one or another classification of knowledge. To establish an essential connection, library classificationists would have to give evidence—in the form of discussions, arguments, and so on—that the same concerns and issues that commanded the interest of the knowledge classificationists, commanded their interest as well.

In reality, there are few, if any, straightforward accounts in which nineteenth-century library classificationists deal in this way with the movement to classify knowledge and the sciences. In fact, there are no really comprehensive and serious general discussions of the classification of knowledge and the sciences among librarians until the writings of E. C. Richardson and H. E. Bliss in the twentieth

---

[5] Cutter listed many of the catalogs that were based on Bacon's scheme in his "Library Catalogues" article published in the 1876 "Special Report" published by the Bureau of Education (Cutter 1876a, pp. 568–622). An analysis of his list can be found in Miksa 1983a, pp. 87–108.

century. And there is no strong obvious connection between the two realms in an actual library classification system until James Duff Brown created the Subject Classification after the turn of the century (Brown 1914).

The lack of an essential connection is evident even in the work of Charles A. Cutter, the early library classificationist who was looked upon by his contemporaries as the most scholarly of those working in the field of library classification. Trying to determine just how close he was to such philosophical concerns has always produced a niggling question at the back of my mind. If Cutter was deeply involved with what was occurring in the classification of knowledge and the sciences, why are there so few examples of that concern in the corpus of his writings? His body of writings is sizable, more than 350 items over a long career. At best there are only two or three articles which delve into the topic of the classification of knowledge in any depth, and two or three brief articles simply do not establish a firm connection. Moreover, even when he did address the matter, Cutter never reached the depth of analysis later associated with Richardson and Bliss; nor did he express anything resembling the less comprehensive concerns of Bradford and Pollard. Further, Cutter's most direct claim of a connection, his statement that the Expansive Classification was based on the idea of evolution, appears to have been tinged by other than strictly philosophical concerns. The ordinary source of this claim comes from the often-quoted address that he gave at the second international conference of librarians in July 1897 (Cutter 1977a). What is less well-known, however, is that he gave essentially the same paper two weeks later at the Institut International de Bibliographie (IIB) conference in August 1897 (Cutter 1897). In the latter instance, however, his main thrust was not merely a description of his system but rather why the founders of the IIB should adopt his classification for their own work. Given what is otherwise a paucity of comments by Cutter on such philosophical concerns, I have become convinced that he made his claim about the connection of his system to the philosophical idea of evolution simply as a way to link his own work to a respectable scientific base so that people like Otlet and LaFontaine, who desired to make scientific and scholarly information available to scholars, might take more interest in it. It is true, of course, that his system is connected to the general idea of evolution, at least in some places. Thus, Cutter was not making a false claim. But the fact is that the connection is neither very obvious nor very strong.

There is another possible source for the claim that there is a firm connection between library classification and the movement to classify knowledge and the sciences. The public discussions which took place in the library press during the first decade of the twentieth century about whether library classifications should be theoretical or practical suggest that librarians were at least aware of the issues raised by the classificationists of knowledge and the sciences. The difficulty here is that what librarians took to be theoretical must not be confused with philosophical discussions about the classification of knowledge and the sciences. For librarians, theoretical appears to have meant primarily that a classification scheme was complex rather than simple, or that an entire scheme or part of a scheme had the imprimatur of current scientists and was worked out in advance of its application rather than being created by non-specialists and expanded as needed.[6]

The DDC and the Classification of Knowledge and the Sciences

The same weak connection can be seen in the relationship of the DDC to the classification of knowledge and the sciences. We are used to hearing about the connection of Dewey's scheme to the work of Francis Bacon, and this tie suggests a connection at least to the beginning of the long-lived movement to classify knowledge. Even Dewey himself claimed this connection with Bacon at the very end of the preface to the 1st edition of his scheme, and he repeated this claim in every introduction he wrote for the system from then until 1927. In 1961 Leo LaMontagne was the first to elaborate on this connection at some length (LaMontagne 1961, pp. 173–88). Even prior to LaMontagne, Eugene Graziano pointed out evidence of a distant connection between the DDC and Hegelianism (Graziano 1959).

Benjamin A. Custer, the editor of the DDC beginning in the 1950s, also touted the philosophical classification of knowledge connections of the system in his introduction to the 16th edition in 1958. Custer attributed the connection to Dewey's serious study of philosophical classifications of knowledge:

> After study of the classification of knowledge as conceived by Aristotle, Bacon, Locke, and other philosophers, and the recently publisht [sic] library classifications of Schwartz and Harris, Dewey decided to use a scheme of arranging books by subject based upon Harris's inversion of the Baconian order of History, Poesy, Philosophy. (DDC 16th edition, p. 5)

This makes good reading, of course, but Edith Scott in her review of the 16th edition was highly skeptical of the connections that Custer claimed. To support her point, she quoted Dewey's introduction to the 7th edition of the DDC published in 1911 (Scott 1960, p. 17):

> The selection and arrangement of the many thousand heads of the classification cannot be explained in detail for want of space. In all the work, philosophical theory and accuracy have been made to yield to practical usefulness. The impossibility of making a satisfactory classification of all knowledge as preserved in books, has been appreciated from the first, and nothing of the kind attempted. Theoretical harmony and exactness has been repeatedly sacrificed to practical requirements. (DDC 7th edition, p. 11)

---

[6] Classification controversy was featured twice in the library press during the first four decades after 1876. The first occurred in the 1880s and focused chiefly on how "minutely" library classification schemes should be divided. The second occurred in the first decade of the twentieth century and focused on whether library classification schemes should be theoretical or practical at base. The first controversy has been investigated in part by Comaromi (1976) and by Miksa (1974, pp. 569–76) as parts of works on other topics. The second has never been investigated so far as I know despite its being a critical element in the library field's development of an ideology concerning "users."

What Scott may not have known is that Dewey's words in the 7th edition were nearly unchanged from those in his 1st edition in 1876[7] and that they matched other statements of his in general tone. By the 2nd edition, Dewey added more of the same:

> Practical utility and economy are the keynotes of the entire system and no mere theoretical refinement has been allowed to modify the scheme if it would detract from its usefulness or add to its running expenses. (DDC 2nd edition, p. 22)

Now, we might conclude that these statements do not mean that Dewey was opposed to contemporaneous philosophical classifications of knowledge, but rather that he took great liberties when he modified them for his own use. They do not say outright that he did not resort to such philosophical classifications of knowledge at all. On the other hand, he seems not to have been much of an aficionado of such systems. In the introduction to the 12th edition, he casts doubt on the usefulness of philosophical classifications of knowledge altogether:

> Tho the importance of classification was recognized, the filosofic sistems proposed wer so difficult fully to understand or apply that not 1 person in 1000 cud uze them practically. Decimal Classification simplicity and even more its Relativ Index hav made this work 10-fold eazier. In recent years, use of the sistem has spred rapidly in all civilized cuntries, meeting success in thousands of different applications. In its simpl form a skoolboy can quikly master it and keep for instant reference not only his books but every note, clipping or pamflet. Almost every profession and occupation has learnd its wonderful laborsaving powers. It is in daily use by miriads of business and professional men who wud never even attempt to understand or uze the old sistems. (DDC 12th edition, p. 10)[8]

Differences in Goals between Library Classification and the Movement to Classify Knowledge and the Sciences

The foregoing discussion has shown that claims of a strong connection between modern library classification in its beginning years and the movement to classify knowledge and the sciences are simply not supportable. At the same time,

---

[7] In the 1st edition, the statement reads as recorded below. The words in italics represent words omitted or changed, e.g., the phrase "the thousand headings" was changed to "many thousand heads," an understandable revision since the DDC had gone well beyond the original three levels of notation used in the first edition. Dewey's omission of words related directly to his work at Amherst—i.e., "the college"—is also understandable for editions later than the first.

"The selection and arrangement of the thousand *headings* of the classification cannot be explained in detail for want of space. In all the work, philosophical theory and accuracy have been made to yield to practical usefulness. The impossibility of making a satisfactory classification of all knowledge as *preserved* in books, has been appreciated from the first, and nothing of the kind attempted. Theoretical harmony and exactness has been repeatedly sacrificed to *the* practical requirements *of the library or to the convenience of the department in the college*." (DDC 1st edition, p. 4)

[8] Here and elsewhere, Dewey's "simplified spelling" is retained.

that there was a connection seems undeniable. Here, an attempt is made to explain how the two movements were related to one another. By examining their respective goals, we shall discover how the goals of library classification were built obliquely, as it were, on the goals of the older and broader movement.

Those engaged in the classification of knowledge and, by the nineteenth century, especially those engaged in the classification of the sciences, appear to have been motivated by a need to explain the value of the rising sciences and the knowledge they produced. Dolby expresses this as the desire on the part of those who were concerned about the classification of the sciences to show how individual areas of scientific endeavor fit into the whole structure of knowledge (Dolby 1979, pp. 173, 187–88).

This motivation will likely sound strange to us today. Throughout much of the twentieth century and particularly since World War II, science has been in a positive and central position in our cultural landscape and belief systems. It has needed neither justifications nor rationalizations to establish its validity. For most of the nineteenth century, however, this was not yet the case. Though the numbers of those involved in scientific endeavors were certainly on the rise, science had not yet gained the pre-eminent social status that it has gained in the present century.

Given this reality, the work of the nineteenth-century classificationists of knowledge and the sciences is more understandable. That work stood as an apologetic. Following a tradition which claimed legitimacy for particular areas of knowledge as they 'fit in' to a whole-cloth portrayal of all knowledge, classificationists of science argued rigorously that this new knowledge, which in reality was dangerously upsetting all accepted conventions of life and the world, was not merely a legitimate part of the whole but was in reality at the core of the whole because it represented the most advanced and established body of human knowledge. Their subsequent production of classifications of the sciences was a way for them to demonstrate this conclusion.

In contrast, library classificationists had a much more practical, though no less exalted, task in view. That more practical task was to make the knowledge found in books accessible to people—in fact, not simply the knowledge found in any books but the knowledge found in the best books by the best authors. The knowledge found in the best books was by definition knowledge produced by the sciences, that is, by those persons who used the best empirical and intellectual methods to discover the secrets of all realms of nature. As such, the works that scholars and scientists produced represented the most authoritative and established representations of humankind's knowledge.

The librarian's task of making such works available had a pedigree that went back to European Enlightenment thinkers who had envisioned the spread of established scientific knowledge among the common people, particularly in the form of encyclopedias, as the hallmark and basis of an enlightened civilization. In the nineteenth century, as a greatly expanded printing industry began to bring the individual works of the greatest thinkers to new mass reading markets, this goal came to be expressed by librarians in terms of the power not simply of reading such works, but of reading them in a structured way—that is, in terms of their relative positions in a classified universe of knowledge.[9]

The goal of librarians was no less important than the goal of the classificationists of knowledge and the sciences, but it had a different purpose. It did not require that library classificationists argue for the legitimacy of the sciences. It required merely that they assume the truth of that argument and proceed onward from that point to their own goal of bringing knowledge in general and especially the established knowledge of humankind (i.e., scientific knowledge) to the public.

At the same time, while it appears that library classificationists simply assumed the legitimacy of the efforts of the knowledge classificationists to justify the value of the sciences and the knowledge the sciences produced, they do not appear to have accepted the supremacy of any one particular portrayal of the realm of knowledge over another. There is little evidence that they found any one scheme more valid than another or even that they found the proliferation of such schemes useful. Rather, they appear simply to have adopted *the utility of the method* used by the classificationists of knowledge and the sciences to portray knowledge. They then proceeded to develop their own schemes on the basis of that method, namely, the portrayal of knowledge in the form of a hierarchical structure.

A hierarchical structure of knowledge categories was deemed the best because it was the belief of those who created such structures that hierarchical relationships among categories of knowledge are "natural"—that is, they reflect the way the human mind produces and uses knowledge and the way such knowledge exists in its various relationships. Further, it was an accepted matter that the categories themselves proceeded hierarchically from general categories to specific, from theoretical categories to applied, from abstract categories to concrete. Its structural elements called subjects consisted of main departments at the top, followed by branches, individual sciences, and so on, finally reaching down to concrete subjects at the lowest levels. In sum, library classificationists, who, in general, were practical in nature, found the hierarchical model to be eminently useful. In their view, the only good way to organize the best books by the best authors on the shelves of a library or in a catalog was in a hierarchically classified structure of subject categories. All other methods were clearly inferior.[10]

---

[9] A brief account of how reading became a force for cultivating minds, particularly when it was directed by the hierarchical relationships among subjects in such a structure, can be found in Miksa 1986, pp. 367–68.

[10] The origin of this approach to knowledge classification can be found in the new mental philosophy that had grown up during the European Enlightenment and in the specific form of mental philosophy called "faculty psychology" which was expounded by a variety of philosophers and educators for much of the nineteenth century. Library classificationists were particularly influenced by this approach to knowledge and the human mind in the form of treatises written by such Scottish common-sense realists as Thomas Reid and Dugald Stewart. An accounting of how this affected librarians' thinking about the nature of subjects can be found in Miksa 1983a, pp. 24–71. A discussion of the relative merits of different kinds of subject access arrangements, and especially of the superiority of classified order, can be found in the same work on pp. 87–123.

Practicality, Dewey's Business Perspective, and the DDC

The commitment of library classificationists to the task of getting the best readings to library patrons in the foregoing manner is only one of the themes that is clearly evident in their writings. Another theme that also emerges is practicality. While there are no extended philosophical essays on knowledge, the structure of knowledge, or other similar issues in library literature, there are long and involved discussions of practicality. This emphasis is especially pronounced in the work of Melvil Dewey, who was the chief promoter of practicality and economy in library work, and it is particularly true of the way he viewed the DDC. When one reads any version of the introductions he wrote for the DDC over the years, his emphasis on practicality not only cannot be missed, but stands out notably. Indeed, we need not go further than this in describing Dewey's general enthusiasm for practicality, because it is apparent everywhere. However, a special aspect of Dewey's merging of practicality with a business perspective merits our attention.

Dewey was not enthusiastic about practicality merely for its own sake. He was also enthusiastic about practicality from the standpoint of business. Dewey pursued library work from nearly the beginning of his career as much from his business sense as from any broader social or educational purpose. In fact, early in his career he had concocted a grand plan for building something akin to a corporate empire while serving such higher endeavors as librarianship and education. The DDC was, from this perspective, simply one more tool in his toolbox of educational and library economy devices which would help him reach his business goal.[11]

Viewing the DDC as the invention of a businessman does not detract from its value. But the fact is the DDC was an entrepreneur's dream, a brilliant invention which would have thrilled anyone who had a business sense and appreciated the value of innovation. It was like an 1870s DOS or Netscape in that it did something that no one else had ever successfully done before—it organized books in a reasonably efficient way in libraries. Dewey made the prototype of the system for the Amherst College library between 1873 and 1876 with help from the Amherst College faculty. In 1876 he published something akin to a "beta" version of the system (i.e., its 1st edition) for use in other libraries. Subsequently, he published a more fully worked-out version at Columbia College in 1885 (its 2nd edition), where he had essentially free help for its development from college library employees. For the next twenty-five years the most difficult task for the system's compilers appears to have been how to determine that magic number of copies to produce which would maximize sales but also keep printing and inventory costs down. In marketing, the scheme practically sold itself. It was certainly user-friendly and easy enough to operate. Moreover, Dewey had such a head start over others attempting to compile a similar system that he had little cause to be concerned about competition. In the end the product "made" the man. Hardly anyone could walk into a library

---

[11] References to Dewey's business ventures occur in nearly all biographical accounts of him chiefly because his business ventures occupied such a large part of his work. Wiegand 1996a, pp. 34–123, the most recent such biographical account, surveys this aspect of his work in some detail. A more limited account of his plans for an overarching library-business organization can be found in Miksa 1983b.

then or now without being aware of his system and, therefore, of him. He always claimed that he made no money from the invention, but this statement is subject to some interpretation. He may well not have accumulated any direct revenues from the system, at least at the start, but certainly he was astute enough to parlay its reputation and value for underwriting other ventures.

Ultimately, one senses much more of Dewey's interest and enthusiasm about the DDC when viewing it in this manner than by attempting to tie it to some deep abiding interest in the philosophical classification of knowledge and the sciences. Dewey was simply not given to philosophical speculations. But he was enormously enthusiastic about a business and practical approach to problem-solving. Looking at the DDC in this way also helps one to understand better his various decisions about it, especially his decision to follow a policy of "integrity of numbers" beginning with the second edition in 1885. His policy of integrity of numbers was in reality his assurance that version 10.0 would work with version 8.0, and that version 8.0 would still work with version 6.0 or version 2.0. It also explains why he was so apologetic about version 2.0 itself when it was published in a 2nd edition in 1885, because that new edition would not easily work with the beta version published in 1876. Obviously, no entrepreneur willingly incorporates serious changes in a successful product unless such changes are absolutely critical to its future.

Dewey's business approach to library classification and the DDC also helps greatly to explain his attitude during the 1890s toward requests from Paul Otlet and Henri LaFontaine of the Institut International de Bibliographie on the one hand and from personnel at the Library of Congress on the other to use and further develop the system. Otlet and LaFontaine simply wanted to build on the DDC as a base. At least that is what they said, and in many respects that is what they did at the start. Their addition of special punctuation signs to the system did not matter as long as the base stayed the same or nearly so. Furthermore, if they were to go out on a limb and expand certain sections while remaining reasonably close to the base that Dewey had created, who would stand to benefit more than Dewey himself. In addition, they were far away and provided access for a relatively new and small segment of the library public—that segment that was only then beginning to make itself known through the documentation movement. Thus, their use of the system did not represent a serious threat to the market represented by the larger and more developed sector of the library movement—established free public libraries. It was a perfect situation. How could Dewey lose?

The Library of Congress on the other hand wanted to change the system fundamentally right from the start. They were candid about this and Dewey's immediate reaction was exactly what one would expect. He refused their request. Here, the issue was more complicated, however. When Library of Congress personnel made their request of Dewey in 1899, the library had already come under the leadership of Herbert Putnam, who was determined to make the Library of Congress the leading library in the United States. It is possible that the reason why Dewey entertained the request at all was that he thought that his friend Putnam could convince the library's relatively new staff to adopt the DDC without alteration. Putnam was not able to do this, however. The long and the short of it was that had the Library of Congress taken this good invention and altered it, the result would have been competition for Dewey, competition which would have likely required basic changes in his own system. But by the late 1890s, his system was too far along in its development to be changed. He was already in the process of defeating one competitor, the *Expansive Classification* by Charles A. Cutter. Why should he encourage another by supplying elements of his own unique work. Thus he turned the request down.[12]

It seems obvious, in sum, that in order to understand the way Dewey himself viewed the DDC, we must first of all contend with Dewey's practical business sense. Dewey obviously used the results of the knowledge classificationists to some degree, and certainly he followed the idea that knowledge when properly classified would be in a hierarchical structure. Thus, it cannot be said that Dewey separated himself from the larger movement to classify knowledge and the sciences. At the same time, it seems clear that Dewey did not consider the concerns of the classificationists of knowledge and the sciences to be the most critical ingredients for making his decimal invention a success. For the latter, Dewey was insistent on practical techniques and results rather than on any philosophical ideas related to knowledge itself.

The foregoing explanation does not fully explain the relationship between the DDC and the UDC over the years. If, in fact, the UDC came to represent something quite different from the DDC, something much more closely allied with concerns about the classification of the sciences, and if these concerns arose as early as 1905 when the first edition of the UDC was published, why then, during the 1920s, did Dewey flirt with the idea of more effectively merging the DDC with the UDC? Only a thorough investigation of all the sources that describe the relationships among the participants in those negotiations will ultimately provide a firm answer to the question. However, based on my research, I offer the following speculation.[13]

Before the 1920s or at least before the end of World War I, although some conflict had arisen between Dewey and the makers of the UDC, it was still generally accepted that the UDC was basically an extension of the DDC. That is why it was regularly called the "Brussels Expansion" of the DDC. Even Dewey welcomed expansions of the DDC by specialists working on the UDC and augmented his own scheme with at least one of them, if not more. At the same time, Dewey could not have missed the implications of the beginnings of scientific specialization, as evidenced by the growth of both the special library and documentation move-

---

[12] LaMontagne 1961, pp. 221–33, was the first to provide an account of the negotiations between Dewey and Library of Congress personnel. However, Scott 1970, pp. 151–227, provides both the fullest description and the most authoritative sources for the matter.

[13] The situation was obviously much more complex than this, a matter quite evident in Rayward 1975, pp. 40–47, 58–68, and 85–111. Despite Rayward's report of "conflict" in the matter, Dewey remained cordial enough with Otlet to accede to the use of some of Otlet's additions to the system in the DDC's 7th edition. A much more decisive break did not occur until the mid-1920s.

ments by the second decade of the twentieth century. Scientific specialization was also the source for much of the pressure on his own system to expand. Moreover, he knew that special fields were using his scheme for purposes other than book classification.[14]

Given at least a cursory sense of such developments, it seems very likely that Dewey saw the prospect of both systems working more closely together as a boon, a matter of extending the influence of both his work and name. Further, he knew Otlet personally and when both Otlet and Dewey's son, Godfrey, suggested pursuing a course of cooperation between the two schemes, the prospect must have seemed a good idea, at least on the surface. What he appears not to have reckoned with was the hard stance that Dorcas Fellows, the editor of the DDC after 1921, would take against the proposal; nor does he appear to have anticipated the friction that would arise between her and Godfrey and between her and newer members of the IIB from Holland and England whose orientation was more scientific than Otlet's. That friction arose when Fellows was directed to bring the two systems into concordance. The correspondence between Fellows and Melvil Dewey for this period portrays Fellows as someone who had taken the task upon herself of reminding Dewey, now advanced in years, of his original practical vision of the system, while at the same time she demeaned the labor and intentions of those who were for the merger. Dewey's responses to her correspondence, often in the form of underlining and interlined remarks, suggest in turn that he was surprised at the need to be reminded of his original goals but happy she had done so. In the end, it appears to be Fellows who convinced Dewey that bringing the two systems into concordance was not in his or the DDC's best interests. As a result, the attempt at cooperation failed and Dewey, now advanced in age, no longer had the personal interest or energy to make it work.[15]

Summary

As this examination demonstrates, the relationship between the two movements—the older movement to classify knowledge and the sciences, and the movement to classify libraries that arose during the late nineteenth century—was not a strong or essential one. Library classificationists took some ideas from the movement to classify knowledge and the sciences—for example, the notion that the classification of knowledge is useful, that it is perhaps the only viable way to view knowledge, and that a good classification of knowledge consists of a hierarchical structure which divides knowledge into categories moving from the abstract down

---

[14] Alternative uses, promoted by Dewey's Library Bureau, focused mainly on the indexing of correspondence (Flanzraich 1993, p. 418).

[15] Comaromi 1976, pp. 297–313, reports the entire sequence of events under the chapter heading, "Foreign Entanglement." However, in many respects he captures only some of the harshness of the exchanges that took place during the period. A heavy concentration of that correspondence can be found in the Melvil Dewey Papers, Box 12, Rare Books and Manuscripts Division, Columbia University Library.

to the concrete. But unlike proponents of the movement to classify knowledge and the sciences, who felt they had to be apologists for the role and status of science in the modern world, library classificationists simply assumed that the conclusions of the knowledge classificationists were both good and even necessary. But apart from that, they had their own task of getting knowledge to library patrons in the form of books. And that task had everything to do with practicality and very little to do with philosophical speculation about knowledge.

PART III

The Relationship of the DDC to Twentieth-Century Library Classification Theory

The previous section argued that when modern library classification began in the nineteenth century, only a weak relationship existed between library classification and the broader and more philosophical movement to classify knowledge and the sciences. That conclusion raises another significant issue, however. If library classification did not arise in some essential way from the seedbed of its more philosophical predecessor, how then does one account for the rise of library classification theory and sophisticated techniques in the present century? Moreover, how does a different origin affect how we view library classification in general and library classification in the form of the DDC in our own day? That such theory and sophisticated techniques now exist is not in question. Library classification theory and technique are everywhere! If, however, they did not come from the nineteenth-century movement to classify knowledge and the sciences, from whence did they come?

The answer to these questions is more complex than might be apparent. First, we must look at what became of the movement to classify knowledge and the sciences. This must be done in order to show that there has not been some sort of delayed relationship of the latter to library classification, a relationship that skipped the first generation of modern library classificationists. With that issue settled, the way will then be clear to examine those ingredients which have provided a basis for twentieth-century library classification theory and technique. Those ingredients include the impact of the rise of the scientific use of information on library classification in the form of the documentation movement, and the discovery within the context of the documentation movement of a new approach to the idea of a subject. The ingredients also include the appearance over time of several key library classification theorists. After examining these matters, we will be able to draw conclusions about the relationship of the DDC to such developments.

The Demise of the Movement to Classify Knowledge and the Sciences

One possible explanation for the rise of twentieth-century library classification theory is to suppose that the library classificationists of the twentieth century somehow discovered the earlier movement to classify knowledge, adopting it as a basis for their own work. In other words, the earlier more philosophical movement constitutes the source of modern library classification ideas, skipping the first generation of library classificationists in so doing.

Two difficulties arise should we choose this explanation. First, only two library classificationists—Ernest Cushing Richardson and Henry Evelyn Bliss—appear to have had personal knowledge of the earlier movement; of these, only Richardson cites earlier classificationists of knowledge extensively by name, though

47

even when he does so he merely lists them without discussing their work specifically. In short, although these two men include more philosophical content in their work than other library classificationists, the ideas of their philosophical predecessors may be found in their writings in only a very general way. One would be hard put to trace specific ideas of particular knowledge classificationists in their work.[1] Secondly, the movement to classify knowledge and the sciences disappeared after the turn of the century, well before substantive theory and sophisticated techniques gained a foothold in library classification. Thus, that earlier movement cannot easily be turned to as a source for modern library classification thought.

The movement to classify knowledge and the sciences ended just after the beginning of the twentieth century, a fact treated by R.G. A. Dolby. Dolby notes that apart from subsequent incidental mention of the knowledge movement in discussions about particular nineteenth-century philosophers who had been active in it, the topic has "become dispersed among the backwaters of intellectual thought"; he concludes that one of the principal reasons for the movement's decline was the "increasing artificiality of the main lines of [its] discussion" (Dolby 1979, pp. 167, 187–88).

One form of artificiality consisted of warnings about the evils of scholarly specialization, a matter no better illustrated than in Robert Flint's 1904 statement that such specializations lead to "sectarianism in science," a sectarianism which, "like sectarianism in religion, is unlovely in itself and baneful in its consequences." The most serious consequence of this attitude for Flint was the loss of a holistic view of the entire fabric of knowledge, and it was principally to correct this loss that Flint promoted the idea that philosophy should chart and adjudicate the boundaries of the individual sciences; it should function, in short, as the science of the sciences—scientia scientiarum (Flint 1904, p. 18). While arguments such as these might have had some value in the abstract, they obviously failed to account for just how far scientific specialization had advanced even by the turn of the century.

Another reason that Dolby gives for the demise of the movement to classify the sciences was that the sciences no longer found this approach necessary for their societal justification. In place of enumerations of the sciences in the context of philosophical arguments about the unity of all knowledge, individual sciences increasingly came to justify themselves by appealing very specifically to the value of the potential applications of their work to society (Dolby 1979, pp. 187–88). While this observation by Dolby is enlightening, it only scratches the surface of a much larger shift occurring at the same time about the relationship of science and technology to society. This shift has replaced nineteenth-century philosophical ideals about the progress of civilization through the mechanism of evolutionary growth of positive (i.e., scientific) knowledge with twentieth-century encomiums to the specific discoveries and values of the scientific and technical community.[2]

---

[1] The work of twentieth-century library classification theorists will be discussed more fully later in this section.

[2] Various aspects of the growth in the status of science in the twentieth century can be found in Cohen (1985), a work ostensibly on the idea of "revolution" as a way to view scientific progress, but especially useful in Parts V and VI on scientific progress in the nineteenth and twentieth centuries; in Dickson (1988), who charts the political context of modern science; and in Holton (1996), who comments on the status of science in the context of attempting to refute current post-modernist views of science.

Dolby's conclusion about the decline of the movement to classify knowledge and the sciences does not apply to all uses of classification, of course. Classification remained a viable activity for the sciences, but became limited to being a method for viewing the particular objects which one or another area of science investigated. Moreover, since the turn of the century, a limited number of fields have also taken up the study of classification itself, although their investigations have tended to be insular, focusing primarily on the techniques and methods of classification rather than on general theory that transcends fields.[3] In the end, only the field of librarianship (including the related field of documentation) continued to focus directly on the classification of all knowledge—in short, on the classification of the elements of the universe of knowledge—as a serious pursuit. In fact, since the 1920s, the classification of knowledge has become more rather than less important in the fields of librarianship and documentation, a development that leads one to ask why this is the case. Did library classificationists and documentalists not realize that the movement to classify knowledge and the sciences had died out?

Now, we could simply conclude with Dolby and others that library classification continues mainly as a practical matter, that it is by and large devoid of substantive intellectual content, and that it continues merely because of inertia in a field in which classification schemes invented late in the nineteenth century continue to be used (Dolby 1979, p. 187; Mayr 1982, pp. 1–48).

I suggest that there is another answer to explain why the classification of knowledge should not only have flourished in the library and documentation fields when it died out elsewhere, but also why it went on to develop a considerable body of theory and techniques based on that theory. I believe that the flourishing of library classification theory and techniques and, ultimately, the relationship of the DDC to this new and enriched approach to library classification, can be explained in terms of the coming together of two key ingredients already mentioned: 1) a critical increase in information production and use among scientists and specialists which began about the turn of the century and which became for certain library classificationists, especially for those in the documentation movement, a laboratory for the discovery of a new approach to, and experience with, the idea of subjects; and 2) the appearance of a small number of theorists in library classification during the first half of the twentieth century who provided justifications for, and a tradition of basic thinking about, library classification problems. Ultimately, the justifications and principles were combined with the new idea of a subject which had been in the making during the same period to form the basis of much of the library classification theory that exists today and which has influenced the character of the DDC.

---

[3] Aside from library and information science, the fields which study classification directly (and their alternative names for this activity) are: Anthropology/Psychology (human mental categorization), Mathematics (set theory), the several Life sciences (Taxonomy), Philosophy, and Statistics (discriminant analysis and clustering). Since the 1960s statistical clustering techniques have become widely used in several fields and a major, though relatively isolated, philosophical discussion regarding it as classification per se was presented by Robert Sokal in *Science* (1974).

Increase in the Production and Use of Information among Scientists and Specialists

The first ingredient in the rise of twentieth-century library classification theory was a critical increase in information production and use among scientists and specialists. This increase began around the turn of the twentieth century and has continued unabated ever since. Most importantly, in the hands of the documentation movement, this increase has become the context for experiencing and developing a new approach to, and view of, the idea of subjects.

The increase in information use had its origin in the rise of scientific and technological specializations at the close of the nineteenth century, along with a corresponding increase in an essentially different, more specialized, kind of literature—scientific and specialist periodicals, proceedings of an increasing number of learned societies, patents, technical reports, and many other kinds of materials which differed from the long-honored, separately published monograph.[4]

It is significant that this increase in the production and use of specialized knowledge is not evident in accounts of the rise of the modern library movement in the United States. These accounts confine themselves largely to the sphere of influence of the American Library Association, the primary agency of the modern library movement. The essential focus of the Association, from its inception in 1876 to well past the turn of the century, was the creation of a general library infrastructure consisting principally of public libraries and agencies which supported public libraries. Because the purpose of these libraries centered on the educational ideal of creating an enlightened, intellectually cultivated citizenry, little concern was expressed about the needs of specialist areas of research.[5] Within the ALA, the principal champions of the needs of specialists were John Shaw Billings, creator of the Index Catalogue of the Surgeon General's Library, and, later, John Cotton Dana, who was responsible for establishing the Special Libraries Association as an entity apart from the ALA. But even these voices were greatly muted. One must look elsewhere for signs of the growing specialist information access area. For example,

---

[4] Most accounts of this change are written from the perspective of individual fields themselves. A brief summary of the change (with appropriate sources) written in the context of a discussion of aspects of research developed from the late nineteenth century to the present, and which includes comments on the idea of a scholar versus the idea of a specialist and on the role of the scholarly monograph, can be found in Miksa (1987) under the heading "Research and Research Patterns."

[5] See the "Epilogue" in Wiegand (1986, pp. 229–36) for an exemplary summary of this orientation through World War I. Wiegand's discussion is written in terms of the ALA motto (concieved by Dewey), "The best reading for the largest number at the least expense" and includes comments both on the nature of mental cultivation goals as well as on the insularity of the ALA in following this course. It would not be until well after the 1920s that other kinds of libraries—for example, school libraries and research libraries—would begin to share substantively in the limelight of the Association's main meetings and stated objectives. It should be noted that the idea of a "public" library in the late nineteenth century focused on any agency that was open to the public, that, in effect, created a "public space" for access to books. Its opposite was a private library. The nomenclature we are so used to today of kinds of libraries (for example, academic, school, special, and public libraries) did not come into widespread use as a way to view the field until nearly two decades had passed in the new century. Shera and Egan (1953, pp. 28–29) portray this lack of concern for scholarly access as a divergence that librarianship took from earlier ideals when, in their words,

other groups created during this period, such as the Medical Library Association (1898), the Bibliographical Society of America (1903), the American Association of Law Libraries (1906), and the Special Libraries Association (1909), had to face this new phenomenon head on.[6]

A serious, even passionate concern about the rise of scholarly and specialist research, as well as a means to bring control and access to the specialized literature associated with that research, was especially evident in Europe during the 1880s and 1890s. Earlier, this interest had been expressed in the publication of the Royal Society's *Catalogue of Scientific Papers* (1867–1902, covering the years 1800–1900), followed by its *International Catalogue of Scientific Literature* (1901–1914). But by the last two decades of the nineteenth century, this concern regularly took the form of bibliographic programs begun as a result of congresses and specialized organizations such as the International Institute of Statistics (1887), the International Congress on the Bibliography of the Mathematical Sciences (1889), the International Colonial Institute (1894), the International Congress of Applied Chemistry (with its International Bureau of Chemical Literature, 1894), and the International Congress of Zoology (with its *Concilium Bibliographicum*, 1895) (Rayward 1975, p. 63). It was in the context of this increased concern and activity that Paul Otlet and Henri La Fontaine began the Institut Internationale de Bibliographie in 1895. The IIB became the single most important organization exercising leadership in the control of and access to specialized literature, and its work is usually viewed as the beginning of the documentation movement, the movement that carried this concern into the twentieth century.

Otlet and La Fontaine began their bibliographical work in 1893 in the International Institute of Sociological Bibliography, but by 1895 had founded the

---

"documentation, librarianship, and archival custodianship were one." They then conclude, "But, very early, subtle forces began to divide those who were interested in these three activities into separate groups which eventually became intolerant of each other." In my view, concluding that these three activities were "one" is questionable because to do so implies the idea that those who operated in the three realms viewed their respective realms as separately organized endeavors. This would be difficult to document, however. At best one might conclude that the three realms were intermixed in the form of people whose work intersected with one another. As a result, if the three areas were never "one" in the way Shera and Egan imply, there cannot have been a point at which they formally diverged. It seems more appropriate to conclude that all three activities began to coalesce individually as more mature bibliographic control traditions were practiced during the nineteenth century enabling the three areas to carve out differing pieces of the bibliographic control pie.

[6] Interest in such needs were sometimes expressed within the confines of smaller ALA sections—for example, in meetings of the college library section—but even there it is difficult to find any serious awareness of the change taking place in scholarly research. When such an awareness did arise, the typical practice was to form a special group. The ALA encouraged such groups to affiliate with it and to meet concurrently with it but did not incorporate their interests into its own program. This became so typical, in fact, that Wiegand remarked almost ruefully that the conference at which the Bibliographical Society of America was formed ended "with another group of special interest librarians organizing outside the ALA" (1986, p. 138). A contrasting picture to that of the ALA's disinterest in the matter can be seen in the creation of the Library of Congress Classification system during the first decade of the twentieth century. That system had to accommodate the scholarly and specialist work that the Library of Congress supported and in doing so embodied some sense of the change that was occurring (Miksa 1984, pp. 25–33; 1983a, pp. 204–11, 398–400).

IIB in the wake of the first International Conference on Bibliography. The energy they expended, including a great amount of travel in order to coordinate the work of many others and in order to convince others to join in their work, was extraordinary. Their primary goal was to create a classified catalog (called the Répertoire Bibliographique Universel) of articles in scientific and scholarly periodicals, proceedings and transactions, monographs and parts of monographs, documents, patents, writings, news reports, brochures, and other non-book materials of all kinds produced by and of use to scholars, scientists, and specialists. They obtained Melvil Dewey's permission not only to base their catalog on the DDC (5th edition, 1894) but to expand the system for their work.[7] The purpose of the Répertoire was to enable scholars to conduct comprehensive bibliographical searches. At first, the founders hoped to sell subscriptions to the cards which they had produced for the catalog, thereby spawning clones of their own catalog, but as time passed, the original catalog became the sole basis of such searches, requests being mailed to them in much the same manner that people today resort to online searching of a distant bibliographic database. Their work during the next quarter of a century was impressive, despite the many hindrances they encountered, for by the early 1920s the Répertoire contained more than twelve million entries and had been used for innumerable searches.[8]

The details of the development of the IIB Repertory, the publication of its *Manuel* and, eventually, the publication of the schedules of its modified version of the DDC in the form of the Classification Décimale Universelle (UDC) need not concern us here. What is important is how Otlet and La Fontaine and their followers approached classification and the impact of that approach on library classification in general.

Otlet and La Fontaine began their venture with the belief that the social sciences had the potential to aid in "discovering a way to surmount the glaring inadequacies of their state." However, since the social science literature was chaotic both in terms of its quality and in terms of control over it, there was little chance that left unattended it could help alleviate social injustices (Rayward 1975, p. 32). The work of the IIB represented a solution to the problem of control that was similar in some respects to Dewey's solution for the general library as implied in his motto "the best reading for the largest number at the least expense." In the case of the IIB, however, the solution was aimed at scholars and specialists and a parallel slogan might have been stated as follows: "providing access to the whole of scholarship for the critical use of the scholarly and specialist community in the most efficient way." To accomplish this goal, Otlet and La Fontaine resorted to the same

---

[7] Rayward (1975) nowhere states explicitly which edition of the DDC Otlet and La Fontaine used, but it seems likely that it was the 5th edition published in 1894. Since the 4th, 5th, and 6th editions contained very few substantive changes in and of themselves, one may conclude that Otlet and La Fontaine were essentially basing their work on the system as it had become filled out in the 3rd edition, published in 1888. Cf. Comaromi 1976, pp. 218–30.

[8] The source of this figure is a 1921 report by W. C. Berwick Sayers, quoted in Rayward (1975, p. 242). The same date and the same total were repeated by A. F. C. Pollard four years later (1925, p. 41).

kinds of practical expedients that Dewey had capitalized on in order to accomplish his own goals—bibliographic cooperation, standardization, and classification.

In using classification to undergird their own solution, Otlet and La Fontaine were no more classification of knowledge theorists than Dewey had been. When they were criticized severely for adopting the DDC, they defended their decision in practical terms. In their view, the DDC and philosophical classifications of knowledge had only one idea in common: both began with the assumption that knowledge, with all of its partitions and elements, consisted of a functional, essential unity, a unity in the same sense that Nature itself was a unity (Rayward 1975, p. 65).[9] Otherwise, for Otlet and La Fontaine, classification was chiefly a matter of practicalities in which a hierarchical enumeration of subject categories was joined to an internationally recognized system of notation in order to provide locations for each subject they encountered in the materials they classified. If anything, the practicality exercised by Otlet and La Fontaine went well beyond that of Dewey's. In working at the task of subdivision, usually through enlisting the aid of specialists, they invented many of the unique notational synthetic devices that came to characterize their version of the DDC.

How did this emphasis on practicality in the work of Otlet and La Fontaine, contribute to the rise of theory in library classification? It is my contention that the experience they gained in attempting to control the literature of specialists helped them forge an essentially new approach to the idea of a subject and to systems of classification which attempted to bring order to such subjects. This experience and the ultimate resolution of the idea of a subject was not comparable to anything prior to Otlet and La Fontaine.

A New View of Subjects and of Classification

The experience of Otlet and La Fontaine in dealing with subjects can be seen in much greater relief by contrasting their application of classification with that of, say, a typical general library, the collections of which were organized by the DDC. Both applications began with the ideal of determining a single place for the subject of each information-bearing entity they classified. In the typical general library situation, however, the number of items found in what we would call smaller, more specific, more discrete, subject categories, were ordinarily few in contrast to the numbers of those kinds of items encompassed by the Répertoire. In addition, because the principal goal of general libraries was to arrange their books in order to educate their patrons in a general way, the topics of such books tended to have well-defined singular themes which could be named with relatively established and recognizable terms.

By way of illustration, a general collection might include a book on the subject called "Coal," or possibly on the subject called "Coal gas." However, it is highly unlikely that such a collection would purposefully acquire a specialist work

---

[9] It may well have been this starting point that caused Otlet to consider the DDC's notation to be an expression of scientific logic, for unlike other notations, its decimal symbols represented elements of the concept of unity.

on the topic "New Developments in the Desulfurization of Hot Coal Gas with Regenerable Metal." If such a work were acquired (for example, by gift), it would be a simple enough matter to include it in a mixed category of similar works at a higher level of a subject hierarchy—for example, under the subject "Coal gas," if such a category existed, or even under the broad topic "Coal." The reason why using a broader general term for the item represented a useful solution is because the numbers of such specialist items were ordinarily few in a general collection and grouping such items together with other similar items instead of in their own unique categories was simply a practical matter with respect to retrieval.[10] In addition, even should such an item be present in such a collection in the form of an article in a general periodical or monographic anthology, or even as part of a single work monograph, it still would not present a problem because, by the 1880s, librarians had already severed analytical work from their cataloging and classification procedures in favor of classificatory categories which applied to all material under a given dominant title. In short, the entire run of a periodical or an entire anthology containing such an item would be classified only as whole single items; classified subject access would be provided at that whole item level only under broader, more simply named categories. Classified access would not be provided for the topics of individual parts, articles, or works of the kind found here in the article on "New Developments in the Desulfurization of Hot Coal Gas with Regenerable Metal."

Another way to state all of this is that as long as the basic bibliographic unit of attention in classification consisted of whole items which dealt principally with more general concepts that had relatively brief and established names, the effort to make a classification scheme with a sufficient number of hierarchical levels to accommodate the subjects of these items was not overly severe or troublesome. This pertained regardless of whether one's basis for determining that singular subject of a document was a scale of concreteness such as Cutter used early on in his subject heading system work, or consisted merely of the vague idea of identifying the central focus or theme of a document, a practice which became more common after the turn of the century.[11]

In contrast to the foregoing practices associated with typical general library classification schemes, Otlet and La Fontaine dealt with specialist articles from the start, and not simply in a few cases but rather in the great majority of cases. Further, they believed that scientists and specialists were only interested in materials at confined (specific) levels. However, dealing with the kinds of information-bearing entities which contained these kinds of subjects led in turn to trenchant procedural problems. For example, because the subjects of specialist treatments could often only be indicated by combinations of more general terms, the problem of subject structure and subdivision became intense. It is a much more complex

---

[10] This presupposes, of course, that neither precision nor speed in identifying documents on a specialist subject of this kind was necessary. Precision and speed were goals that were shaped by the needs of scholars and specialists that the documentation movement served. As such, they would not become part of the everyday vocabulary of retrieval until after World War II.

[11] Cutter's use of the term "specific" to refer to a scale of concreteness is discussed at length in Miksa 1983a.

problem to locate a hierarchical place for a topic indicated by phraseology such as "New Developments in the Desulfurization of Hot Coal Gas with Regenerable Metal" than it is to do the same for a treatise wholly on "Coal" or even wholly on "Coal Gas." Again, the deeper one's hierarchy of subjects was extended to accommodate such topics, the more options one had in ordering the divisions of such topics in a useful way. This realization forced classifiers to make choices as to the kinds of subdivisions they needed and the order in which they listed them. Still again, the greater the number of complex formulations of topics one encountered, the more likely one was to find elements of the formulations in different subject statements containing the same terms. In the example topic here, "desulfurization" and "regenerable metal" are terms that might appear in more than one place in a scheme dealing with topics of this level of discreteness. Finally, simply determining how to speak of such subjects was a problem. Does one refer to them as we have been doing here as subjects of narrower rather than broader focus, or are they subjects that are smaller rather than larger? Or are they subjects of greater intension but lesser extension as opposed to subjects of greater extension and lesser intension?[12]

Otlet and La Fontaine provided an example of the discovery of this new view of subjects. Others who expressed the new idea of subjects (their complexity, their problematic position in a classificatory hierarchy) and who were also connected to the IIB, included Fritz Donker Duyvis, A. F. C. Pollard, and Samuel Bradford. Bradford and Pollard, the organizers of the British Society for International Bibliography (the British arm of the IIB) have left a particularly rich collection of writings which wonderfully illustrate the same discovery.

Like Otlet and La Fontaine, Bradford and Pollard focused in their writings on discrete complex subjects found far down the hierarchical chain. Their goal was to fulfill the needs of specialists and scientists, who wanted to be able to find specific information on subjects in their areas of research without having to browse though subject sections of lesser specificity. Because of their own scientific education, Bradford and Pollard were insistent on the urgency of the matter. Their sense of urgency, however, did not arise from the philosophical concerns associated with Otlet and La Fontaine, but from the fact that they belonged to the new generation of scientist-librarians who were also information specialists.

Bradford and Pollard's writings especially illustrate the complex relationships of the kinds of subjects they viewed. How, in fact, could one structure a classification scheme in such a way as to accommodate such topics? How could one integrate these subjects into a schema, given the complexity of their names, in such a way as to be able to find them in a specific location within the scheme when needed?[13]

---

[12] W. C. Berwick Sayers introduced the ideas of intension and extension, which he got from Aristotelian logic, to library work. Subsequently, the ideas were adopted by Ranganathan and used widely among his followers.

[13] It should be noted that Bradford and Pollard continued to think of such subjects as occupying specific locations within classificatory systems, despite the facet-like nature of the UDC with which they worked. Thus, their use of faceted structure was for them simply a technique which would allow the useful plotting of all such subjects into individual locations in a classification system.

Bradford and Pollard did not approach this discovery of a new idea of subjects as classification theorists. Like Otlet and La Fontaine before them, and like Melvil Dewey before that, they were practical men. Their practicality was not that of businessmen, however, nor of social reformers, but of scientist-engineers. Their work ultimately contributed to the growth of theoretical library classification, but it should not be confused with a straightforward determination to generate library classification theory. In short, they were no more classification theorists than those who had preceded them.

The Rise of Library Classification Theory

The enormous growth in the production and use of scientific and specialist information and the discovery of a new view of the idea of a subject in relationship to it was one important ingredient in the creation of modern library classification theory and technique. This growth was paralleled by a second ingredient: the appearance of a group of persons who began a tradition of discussion and theorizing about library classification issues. These people were few in number until the late 1940s when full-blown discussion of theoretical library classification ensued. I will discuss four such theorists here—Ernest Cushing Richardson, Henry Evelyn Bliss, William C. Berwick Sayers, and S. R. Ranganathan.[14] In many respects one will find the seeds of nearly all subsequent library classificatory principles embedded in the work of these four men.

Ernest Cushing Richardson

The first of the four classification theorists was Ernest Cushing Richardson (1860–1939). Richardson had been a significant participant in the creation of the "modern library" during the last two decades of the nineteenth century, and he continued actively working in the library movement as it gained significant maturity during the first third of the twentieth century.[15] He spent most of his professional

---

[14] Doubtless, some would argue that there were others besides these four who should be included in this list of twentieth-century library classification theory pioneers. It is tempting, for example, to include James Duff Brown, the creator of the *Subject Classification* during the late 1890s and the first decade of the twentieth century. He is not included here chiefly because in a manner similar to Charles A. Cutter, he focused on the details of making an actual classification scheme rather than on any extended discussion of classification theory issues themselves. This was the case despite the fact that his own scheme, called the *Subject Classification*, was unique in breaking away from other contemporaneous systems in how it sequenced topics. Of the four persons listed here, only one, Sayers, did not create a classification system of his own sometime during his life that illustrated his principles. He was a teacher of classification practice and principles and his importance resided not only in how he conceptualized the library classification enterprise but also in the influence he had on the fourth person listed here, S. R. Ranganathan.

[15] The phrase "modern library" is used here in the sense in which it is found in Miksa (1996) as the "public space" library which was invented in the latter part of the nineteenth century. This manifestation of the library, which virtually all persons involved in libraries in the present day consider standard for libraries and which contrasts with the "private space" library which preceded it, solved the problem of the affordability of libraries for citizens at large. As such, the modern library consisted of a social organization open to the public and supported by public funds. Its chief cultural legacy has been to help reinforce the standard, now widely accepted, that information accessibility is a necessary component of modern civilization and virtually a "right" of the members of such a civilization.

years as the librarian of Princeton University (1890–1925) where, among other things, he created a shelf classification scheme for its collections. Afterwards he also served as an influential consultant in the bibliographical work of the Library of Congress. The central thrust of Richardson's professional work was to help bind the task of bibliography, especially national bibliography, to the work of the modern library. But in many respects the most significant specific contribution he made to the modern library movement, and especially to library classification, was his role as one of the modern library movement's most perceptive intellectuals.

Richardson's influence as an intellectual appeared early in his professional career. He appears to have been the person chiefly responsible for influencing Melvil Dewey to include "bibliography" as one of two major defining parts of the curriculum of Dewey's pioneering School of Library Economy at Columbia College. Furthermore, in lectures given before the school's first class in 1887, Richardson argued persuasively that a classified approach to knowledge was the natural framework for library bibliography and reference work. In one of his lectures he spoke of the librarian as "cyclopedic" in order to focus on the professional reference worker as one who was able to call up from memory the best authors and their best writings because he or she could plot them within a classificatory framework of knowledge. This approach to the librarian's task ultimately facilitated choosing the best books for individual readers in particular reference situations.[16]

Richardson turned his attention more formally to library classification when he returned to Dewey's library school (then at the New York State Library in Albany, New York) to give a series of special lectures in 1900. His lectures were published in 1901 under the title *Classification, Theoretical and Practical*, and in subsequent editions in 1912 and 1930 he added an extensive bibliography of classifications of knowledge and of books pertinent to library classification.

Richardson's lectures on classification represented the first concerted attempt by an American librarian to identify basic philosophical principles underlying library classification. Of the various principles he identified, three were especially important. First, he concluded in a kind of correspondence theory of categorization that while the classification of knowledge necessarily meant the identification and arrangement of ideas as categories, such ideas must in actuality correspond to "things"—that is, to objects and, ideally, to all objects, real or imagined, material or immaterial, in existence. In short, library classification had all existence in its purview.

---

[16] Richardson's role in influencing the curriculum of the Columbia School of Library Economy is discussed in Miksa (1986, pp. 374–75). At one point he expressed the idea that the librarian was to be an "encyclopedia," but this was certainly strange wording. Richardson's intention in using the phraseology was not to portray the librarian as one filled with facts and random bits of information, an association with the idea of being encyclopedic that is especially common to the twentieth rather than to the nineteenth century. Rather, he was emphasizing a quality of mind, a sense of the wholeness of the universe of knowledge that implied having a structural framework for the universe of knowledge, a framework commonly associated with the original idea of an encyclopedia as it arose out of the encyclopedist movement of the seventeenth and eighteenth centuries. The lectures given by Richardson and others have been transcribed and are described in Miksa (1988) but have not yet been published for general use.

Second, he concluded that any listing of such objects as categories must seek to place them in the order in which they naturally arose, either in nature or in the mind, the classificatory order becoming thereby a reflection of that "natural" order. Although he concluded that there were a variety of potential orders depending on the categories with which one was dealing—for example, logical, chronological or genealogical (i.e., including evolutionary) orders—all such orders properly reflected a movement from simple to complex. In short, knowledge classification was best done by attempting to reflect the order of "things" in their natural progression from the less complex to the more complex.

The third principle of library classification that Richardson elicited, unlike the first two which dealt with the categorical structure of knowledge classification, focused on the application of structures of knowledge to library work. His conclusions reflect the serious difficulty that library classificationists have with theory. Having erected a notable theoretical structure of ideas related to the categories which librarians needed to use, Richardson, like so many of his contemporaries, then felt it necessary to point out that practical considerations in applying such a structure to the arrangement of books were to take precedence over any consistent rendition of such a theoretical structure. In short, theoretical considerations were to be set aside because of pragmatic needs even when the result produced inconsistencies in the structure of knowledge itself.

Henry Evelyn Bliss

Henry Evelyn Bliss (1870–1955), the second of the four library classification theorists I shall discuss here, set a personal goal early in life of mastering human knowledge. However, in the words of one biographer, when he became frustrated with the prospects of reaching his goal, he concluded that "the only way to achieve even limited mastery of knowledge was to determine the relationships between subjects, the key to which lay in devising a proper classification scheme for knowledge as an organized whole" (Anderson 1978, p. 37). In short, classification work for him grew out of a larger and more personal quest to comprehend the nature of the universe of knowledge.

Bliss pursued his classification work throughout his long tenure (1890–1940) as a librarian at the City College of New York (CCNY) and afterwards during his retirement years. Early in his career he became personally acquainted with two notable classificationists, Charles A. Cutter and Ernest C. Richardson. Throughout his earlier CCNY years he also regularly published articles on knowledge and the sciences and on contemporary library classification systems. He assessed the strengths and weaknesses of the latter but found all such schemes wanting chiefly because they tended to pay little attention to the organization and growth of the modern sciences.

Bliss eventually created his own library classification system, the *Bibliographic Classification* (1st ed., 1935; 2nd ed., 1940–1953), but before doing so he published two important monographs which brought together his views on library classification practices and theory. One work, *The Organization of Knowledge in Libraries* (1933), brought together in one place his many assessments of contemporary library classification systems. In the other work, *The Organization of Knowledge and the System of the Sciences* (1929), with a preface by the philosopher John Dewey, Bliss described his philosophical justification for library classification and gave an elaborate explanation of the general goals and methods of library classification. As such, this work served as the most comprehensive theoretical basis for library classification up to that time.

The first part of Bliss's 1929 work provided a rationale for library classification. He claimed that library classification represented the culminating element of society's efforts to provide "organization." The purpose of organization was to harness society's efforts and resources to combat the ills of modern society. In Bliss's view, organization was a social phenomenon, one important element of which was the social process of organizing knowledge. Organizing knowledge itself advanced in stages and included not only the organization of knowledge generally and in formal educational ventures, but more specifically and ultimately the organization of knowledge in libraries. And organizing knowledge in libraries specifically meant its classification.

Bliss's rationale for library classification represented a striking break from typical nineteenth-century reasoning. In the earlier period, the importance of library classification was typically viewed in terms of how it helped people improve their minds and character. In contrast, Bliss (along with Otlet, La Fontaine, and others in the documentation movement who served scientists, scholars, and others) emphasized the twentieth-century theme of what people could do instrumentally with knowledge made available through classification. More specifically, the importance of library classification was viewed in terms of its instrumentality in putting knowledge to work to solve society's problems.[17]

Bliss defined classification in the next section of his work. He laid out its general procedural principles (viewed mainly in terms of creating a structure of logical categories), and provided a philosophical basis for knowledge classification which fulfilled the culminating goal of "organization" in the first section of the work. His rendition of a philosophical basis for the task was especially significant for it greatly extended and sharpened Richardson's approach to the matter. Bliss began with Richardson's assertion that knowledge classification is first of all the

---

[17] This focus is also found in the extraordinary work by William S. Learned, *The American Public Library and the Diffusion of Knowledge* (1924), which is one of the earliest appearances of this theme. The focus on instrumentality and solving social problems also appears to have represented Bliss's own attempt to come to grips with the enormous shift in cultural outlook represented by World War I and the shaking of nineteenth-century ideas about the inevitability of automatic social progress. The 1920s were years of enormous contrasts in this respect—the growth of wealth among the privileged, dissonance in the arts, social unrest expressed in the form of the age of excess, the fear of depersonalization as a result of industrialization, all counterbalanced by a growing belief that science and technology could solve society's ills. Of course, not everyone was equally hopeful about solutions to such problems. Aldous Huxley's *Brave New World* (1932) gives a much more pessimistic view of things, as does Oswald Spengler in his *Decline of the West* (1926).

classification of the "things" or "objects" of existence ordered as categories. Bliss established this argument by putting forth an elaborate philosophy of realism. He posited an objective world apart from humankind's observation of it and asserted that not only are "objects" of this world discovered progressively, but also that individual minds together constitute a unity in perceiving such objects. He used the term *Nature* to refer to the realm of objects which exist apart from the human mind, and the term *World* to refer to Nature *together with* the realm of objects made up of human ideas and artifacts. He concluded, however, that Nature is more systematic and, therefore, "more coherently knowable" (p. 175) through science than the World generally. This was the case because, despite the fact that Nature is fundamentally evolving and developing, it maintains stable, persistent, and constant relations among its elements which give rise to a discoverable order. Since it is this order of "real relations" which science is in the process of discovering, it is this order that should provide the basis for a classification of knowledge useful to libraries (pp. 185–86).

In a third section Bliss takes his assertion that nature's order provides the best basis for a classification of knowledge one step further by showing how nature's order might be ascertained. His solution was to demonstrate that the order of the sciences themselves—including their character as fundamental sciences, branches of the fundamental sciences, subsciences, composite sciences, new sciences, etc., and the position of individual sciences among all sciences—reflected the order of Nature. Further, he asserted that this order had gained a consensus among scientists and knowledge classificationists, a consensus that was sufficient enough to allow them to trust it as a relatively reliable and permanent structure of knowledge categories. In short, the most appropriate classification of knowledge could and should be a structure of the sciences themselves because such a structure was permanent enough to be useful for years to come.

Thereafter in this section, Bliss carefully examined various possible ways of rendering the order of the individual sciences. His discussion of how this might be accomplished included a principle of arrangement which he called "gradation by specialty." Gradation by specialty meant "the principle by which the several sciences and studies, distinguished by their conceptual scope and their relations to the real order of nature, are arranged in serial order from the most general to the most special" (p. 217).[18]

---

[18] This entire rendering of Bliss's ideas in *The Organization of Knowledge and the System of the Sciences* is necessarily brief and therefore overly simplistic, with only a minimum number of actual page citations being provided. His argument is very detailed and progresses steadily through his work in a such a well argued way that only a close reading of these three sections will help one to appreciate the sophistication he brought to the task. He also included a fourth section entitled "A Historical Survey of Systems of Knowledge" in which he traced systems of knowledge put forth by individuals from Plato to the twentieth century. But this section is not essential for understanding his main arguments because he appears to have used it primarily as a way to demonstrate the correlation he had already drawn between the order of nature and the order of the sciences on the one hand and the principle of "gradation by specialty" on the other hand. The final chapter in the work is a useful summary of his entire argument, however, and also serves as something of a preachment on the necessity of his findings for the task of knowledge classification.

William Charles Berwick Sayers

W. C. Berwick Sayers (1881–1960), the third of the library classificationists to be discussed here, was one of the most influential librarians in England during the first half of the twentieth century. He spent nearly his entire professional career, which began in the late 1890s, in public libraries, chiefly in the Croyden, England, where he was deputy librarian under Louis Stanley Jast from 1904 to 1915 and the chief librarian from 1915 to 1947. His tenure at Croyden was significant, for under his leadership the Croyden Public Library became one of the leading centers of cutting-edge modern librarianship. He also exercised enormous influence on the field through his extensive leadership in library association work and through his extensive and influential role as an author. His writings included countless periodical articles on a wide range of library and other topics. He edited library periodicals (including the *Library World,* established by James Duff Brown) and no less than four editions of Brown's influential *Manual of Library Economy.* He authored a series of monographs on various topics, including four important works on classification. Finally, he exercised considerable influence through his role as an examiner for the Library Association beginning in 1910, and through his many years as a lecturer on classification at the University of London School of Librarianship from its beginning in 1919.

Sayers' initial approach to library classification can be characterized as his own attempt to understand library classification. He first encountered the topic soon after becoming a library assistant when he found it necessary to use the library classifications of the time. But this early experience plunged him "immediately into complete bewilderment." James Duff Brown's brief and practical *Manual of Library Classification and Shelf Arrangement* (1898) helped explain some aspects of the topic for him, but it also made him want to know "by what impulsion philosophers and librarians designed so many and diverse systems; what reasoning made them prefer this or that scheme." He soon discovered E. C. Richardson's *Classification: Theoretical and Practical* (1901) which he read so often he nearly memorized it. Then, in 1905, Sayers attended a series of lectures on classification delivered by Louis Stanley Jast (his superior at Croyton) at the London School of Economics. Sayers often referred to these lectures as a turning point because they convinced him that "this apparently dry-as-dust study could be one of the most fascinating, suggestive and educative within the province of the librarian" (Sayers 1915–1916, pp. 13–14).

Sayers' initial writings on library classification consisted of articles in which he examined Melvil Dewey's *Decimal Classification* and James Duff Brown's *Subject Classification.* He also wrote a thesis on Charles Cutter's *Expansive Classification* which was accepted for his diploma from the Library Association. In 1908 he published an article entitled "The Principles of Classification" in *The Library Assistant,* which was also the first number in the Library Assistants Association series. He reissued it in expanded form in 1912 under the title *The Grammar of Classification* and thereafter it went through two more editions (1924, 1935). In its twelve pages he attempted to state the essence of library classification in thirty-two brief, crisply written paragraphs that were similar to a set of logical propositions or assertions. He had come to his conclusions by studying the library classification systems mentioned earlier. His second work, *Canons of Classification* (1915–16) contained the bulk of his observations about those systems. It also included an additional chapter on the *Library of Congress Classification*.[19] In both works and thereafter Sayers used the term "canons" as a synonym for "basic principles."

A second stage in Sayers' approach to library classification overlapped with the first but then continued afterward as the central focus of his work. This stage, which began when he became an examiner for the Library Association in 1910, featured Sayers not as someone who was struggling to understand library classification but rather as a teacher who was attempting to explain library classification to students. He devised a set of correspondence lessons for library students to help them study the topic of classification before taking their examinations for certification by the Association. He thereafter published his lessons in *An Introduction to Library Classification, with Readings, Questions and Examination Papers* (1918), a work that went through another eight editions by 1954. In each of the editions, Sayers adjusted, expanded, rearranged, and rewrote his explanations so as to state them ever more clearly.

In the first edition of his *Introduction* Sayers expanded his library classification principles from his earlier thirty-two paragraphs to six chapters of more than seventy pages. In a second section of this work he included a chapter about readings related to classification, a chapter of historical background in which he described the classification work of Sir Francis Bacon, Jacques-Charles Brunet, and others as this work applied to libraries, and two chapters of comments on the four widely used library classification schemes he had already examined more extensively in his *Canons*. Finally, he also included two chapters on "practical" classification issues, although it seems obvious that, despite their inclusion, the problem of practical application did not occupy him with the same kind of troubled concern one found in Richardson and others.[20] He was more centrally interested in explaining the logical and systematic principles of library classification and did not see the practical application of library classification as antithetical to library classification logic. Sayers followed his *Introduction* with *A Manual of Classification for Librarians & Bibliographers* (1926), a work in which he brought together all of his thoughts on library classification. He revised this work in two more editions by 1955, and since his death in 1960 it has been issued by others in three more editions.

---

[19] The Library of Congress began issuing what often amounted to "preliminary" editions of its system beginning in 1898. It was not until 1911, however, that most of its schedules even in this preliminary state had been published—namely, schedules (although not all complete) for all areas of knowledge except Language, Literature, and Law. Thus, Sayers could not easily have examined the system before that time. See Miksa (1984) for details concerning the publishing history of this system.

[20] Later editions of this work included a fourth section which extended the "practical" classification issues, by explaining characteristics of the most general knowledge divisions or areas (e.g., religion, social sciences, fine and recreative arts, etc.) found in typical library classification systems. Still, even in doing this, Sayers did not apologize for the practical nature of classification as other writers had.

The essential nature of Sayers' overall contribution to library classification theory can best be described in terms of three interrelated themes. First there was library classification, which Sayers described in a straightforward manner as the arrangement of library books into classes and the arrangement of these classes of books into a system of classes based on class relationships. Moreover, classification was not some *sui generis* activity of librarians. Rather, library classification was neither more nor less than an instance of classification in general. In short, all classification constituted a unified activity of which library classification was merely a particular expression. This was an important starting point for Sayers, not only because it identified an origin or source of classificatory ideas and procedures for library classification, but also because it showed that library classification was indebted to and based on ideas and principles which extended beyond the library field.

Of course, identifying the origin of library classification in this way would not have been very significant if its more general origins also consisted of some idiosyncratic, far-fetched, or eccentric activity in its own right. Quite to the contrary, however, the origin of classification in general for Sayers was nothing short of that magnificent attribute of humankind that sets it apart from lower forms of animals, namely, the everyday human mental activity of distinguishing and naming classes and of analyzing the relationships among ideas, a process that in its refined form is basic to scientific discovery. In sum, the origin of classification was the activity of the mind's reasoning power to generate and use concepts.

Second, given the unity of origin for all classificatory activity and for library classification in particular, as well as the focus in classification on how concepts are derived and related to one another, Sayers also thus had his source for the principles and procedures that must be followed for the activity of classifying. That ready source consisted of the laws of thought that had been discovered and explicated over hundreds of years going back to the time of Aristotle, laws that could be found described in the twentieth-century field called logic. Logic deals with how the mind reasons and, more specifically, with how it reasons rightly. And within that context, logic also deals with the process of classification, for in this approach classification is an integral part of the logical reasoning process. Given the high position of classification within the science of logic, it became evident that should one wish to become a good library classifier or classificationist, one had to understand those laws of thought and, especially, those elements of the laws which the best writers in logic had concluded were basic to the classificatory process.

Third, although Sayers readily pointed out the source of classification in logic and, in fact, listed texts and parts of texts in logic which his students might read to become familiar with ideas relevant to library classification, he did not also subscribe to the idea that students had first to become experts in logic and all of its various facets before becoming library classificationists. Rather, he suggested such writings so that his students might find in greater detail what he himself had written in summary fashion—descriptions of how the mind works so that library classification can be approached with some sort of reasonably consistent methodology. He readily pointed out that these texts were the basis for his writings about the theory of classification. In reality, what he wrote could be more accurately described as an abstraction or a condensation of ideas taken from such sources. I do not say this to detract from Sayers' work. His ability to cull a set of principles or canons of library classification from these longer, more detailed discussions of logic and of how the mind works was truly remarkable. Further, he was so passionate about the connection between the laws of thought as delineated in the fields of logic and library classification that he inspired more than a generation of students to follow his lead, to take the connection for granted and incorporate it into their thinking. In fact, it may be reasonably concluded that his teaching and texts on the topic contributed more than any other source to equating the library classificatory process not only with logic, but also, and more specifically, with that branch of logic that ultimately had its origins in an Aristotelian approach to categorization. The most striking correlation that he made in this respect was to adopt definitions, relationships, and operations that arose from Aristotle's five predicables—i.e., genus, species, difference, property, and accident. Sayers' use of these ideas included describing the process of classificatory division as that of "genus et differentiam," defining a class monothetically as consisting of members which all had the same attribute or attributes (i.e., the same "difference"), describing the relationship between genus and species in terms of intension/connotation and extension/denotation, and differentiating between natural and artificial classification on the basis of how accidental attributes are employed in the classificatory process.[21]

The importance of Sayers' identifying and connecting the principles of library classification with logic, and more specifically with the language and ideas of Aristotelian logic, and his passionate and influential teaching of this connection to more than a generation of students, cannot be overestimated. For the first time library classification had been given a clear and strong methodology that was both understandable and teachable. The methodology clearly insisted that library classification was nothing if it wasn't logical, where logical referred to classical approaches to definition and the delineating of classes and class hierarchies. Further, by clearly identifying library classification with logic defined in this way, Sayers was able to structure the methodology of library classification in terms of canons and axioms. The emphasis is important, especially for its impact on the thinking of Sayers' most notable student, S. R. Ranganathan.

Sayers was eminently successful in his approach to library classification theory, but his work also included a clear sense of irony and had a negative side effect. Irony resided in the very sources that he used for his logical principles. Although as the years passed Sayers cited a number of writers on logic that library classification might make use of, the one writer he most often cited and from whom

---

[21] It would also appear that Sayers' differentiation between knowledge classification and bibliographical classification was affected by the same source of ideas. Sayers considered knowledge classification to be a pure set of subject categories which had been arrived at by means of the "genus et differentiam" process, whereas a bibliographical or library classification consisted of a knowledge classification to which division based on accidental characteristics had been added, with the latter being necessitated by those attributes of books other than their ideas.

he appears to have taken most of his ideas was William Stanley Jevons. Jevons was a brilliant logician and economist who died at a young age in 1882, but whose works on logic—particularly his *Elementary Lessons in Logic, Inductive and Deductive*, first published in 1870, and his *Principles of Science, a Treatise on Logic and Scientific Method* (with an extensive chapter on classification), first published in 1874— continued to be published and used widely well into the twentieth century. Using Jevons was ironic because it was Jevons who in his *Principles of Science* had pronounced library classification by subject to be a "logical absurdity" (Jevons 1877, p. 715). Sayers more than once drew attention to this pronouncement (following J. D. Brown and W. S. Jast in doing so) and there can be little doubt that he dedicated his own work in great part to proving Jevons wrong, to showing that library classification by subject was indeed a logical process.

The negative side effect of Sayers' approach to library classification theory was that it effectively narrowed the base for investigating the process of library classification. This is the case because the "logic" that Sayers identified as basic to library classification in reality provided only a relatively narrow range of approaches to the thinking process and, therefore, to the process of categorization useful for information-bearing entities. For example, by the time Sayers was writing his principles, investigations by George Boole, Augustus DeMorgan, and others, when combined with the work of the German mathematician Georg Cantor, were already yielding the entire field of axiomatic set theory. But Sayers never included such developments in his considerations of library classification theory. This was the case even though Jevons himself had written about Boole. Nor did Sayers take into account later work in statistics (for example, discriminant analysis and clustering) and many-valued logics. And this is to say nothing at all about classification principles and work which have arisen in linguistics, psychology, and anthropology under the general label of the processes of human mental categorization. When one follows out the implications of alternative methodologies such as these, the very idea of classificatory principles will expand significantly from its narrow base in the kind of logic Sayers saw as essential to it. An expansion of that kind subsequently did begin to take place, but only during the last decade of Sayers' life.

Shiyali Ramamrita Ranganathan

S. R. Ranganathan (1892–1972), the last of the four library classification theorists to be discussed here, has been by far the most significant contributor to library classification theory and technique in the twentieth century. Ranganathan was born in the Brahmin community at Shiyali in the Tanjur District of Madras State, India, and learned Hindu thought and philosophy from an early age. He attended Madras Christian College from 1909 to 1916, eventually completing a master's degree in mathematics there. Ranganathan taught mathematics and physics at various colleges from 1917 to 1921, and algebra, trigonometry, and statistics from 1921 to 1923 as an assistant professor at Presidency College in Madras. During the same seven-year period he also began his scholarly career by publishing articles in the area of correspondence theory in the general realm of Theory of Numbers.

Ranganathan's professional life changed abruptly in 1924 when he accepted an appointment to the post of librarian at the University of Madras. The terms of the appointment included an initial year of training in England where, from 1924 to 1925, he studied under W. C. Berwick Sayers at University College of the University of London, completed an internship at the Croyden Public Library under Sayers' direction, and visited a large number of libraries in England in order to observe their work and techniques.[22]

Ranganathan's subsequent professional career can be usefully divided into two parts which differ chiefly in the context and outlook of his work. The first part extended to 1947 and included his work as the chief librarian at the University of Madras (1924–1944) and at Banaras Hindu University (1945–1947). During this period Ranganathan's attention was focused for the most part on the libraries which he administered and on Indian library needs. With respect to the latter, he became for all intents and purposes the father of the modern library movement in India, nearly single-handedly creating workable library goals, tools, and organizations for Indian libraries of all kinds, instituting the beginning of Indian library education, helping to organize and promote library association work, and both promoting and contributing heavily to the publications of the fledgling field of Indian librarianship. During this period he also created and began the explication of nearly all of his most fundamental ideas related to librarianship.[23]

The second part of Ranganathan's professional career began with his appointment as a professor of library science at the University of Delhi (1947–1955) and extended to the establishment of the Documentation and Research Training Center in Bangalore (1962) where he served until his death in 1972. This period of his career differed greatly from the first because rather than focusing on the library needs of India or on university libraries which he directed, his concerns and thinking became much more comprehensive and universal in scope. It was during this period that Ranganathan encountered the documentation field and participated extensively in the activities of the Fédération Internationale de Documentation (FID), the successor organization to

---

[22] The events which led to his appointment appear to have been complex and most likely included, among other things, Ranganathan's role in promoting Indian needs not met by the British colonial administration, an awareness by officials of Ranganathan's love of books and learning and his use of these in his teaching, and Ranganathan's own need for a better salary (the librarian's position paid more than he was earning as a teacher). Ranganathan was not totally convinced that making the change was the right thing to do and even attempted to back out of the arrangement once it had been made. But he persisted nonetheless, and eventually reshaped the entire idea of librarianship in India. His stay in England was apparently from the fall of 1924 to about the mid-summer of 1925.

[23] All such ideas could ultimately be found explicitly or implicitly in his 1931 treatise, *The Five Laws of Library Science,* but he developed major areas of those ideas in one or more separate works—for example, on library catalogs (1934, 1938), on library administration (1935), on library reference service (1940), and on school and college libraries (1942). His formal works on classification during this period included two editions of the *Colon Classification* (1933, 1939), the first edition of his formal treatise on library classification theory, *Prolegomena to Library Classification* (1937), and two works which both explained and expanded his classificatory ideas, *Library Classification: Fundamentals and Procedure* (1944) and *Elements of Library Classification* (1945).

the IIB.[24] He also traveled on numerous lecture and teaching tours in England, the United States, and elsewhere during this period, and strenuously served and promoted the interests of international cooperative work in classification.

Ranganathan's "discovery" of documentation is especially significant for it caused him to reorient his classification work. Previously he had applied his classificatory ideas to what by any measure were small libraries (relatively speaking) which served general populations of users. In contrast, documentation focused on the much more intense arena of the information needs of scientists, engineers, and scholars from numerous other individual modern fields. Ranganathan's response to the needs of the latter was to add to his classificatory work a larger, more complex enterprise called "depth classification," a realm of more sophisticated classificatory structures that dealt with "micro-thought" and "micro-documents."[25]

The most striking single characteristic of all of Ranganathan's work and especially of his work in library classification is that he treated library classification as a single unified structure of ideas which flowed from a cohesive set of basic principles. His propensity to approach librarianship this way likely resulted from his exposure to Hindu thought and mysticism on the one hand and from his study of science and, especially, mathematics on the other hand. Both sources placed great credence in the existence of fundamental principles or laws which lay at the center of life and all existence. To discover these laws in operation was to discover the very nature and order of things, an order based on principles which are eternal, unchanging, and all-encompassing. There is virtually no area of Ranganathan's work and personal life in which this quest for discovering the inner or essential order behind the visible world is absent.

Modern science and Hindu thought provided general sources for Ranganathan's classificatory ideas, both contributing many specific analogies for his ideas. Of these two realms, however, it was mathematics which supplied the base for Ranganathan's most powerful classificatory ideas and, in turn, his most enduring contributions to library classification theory. It should be noted at the start, however, that Ranganathan invoked mathematical concepts in his classification work primarily in an analogical rather than in an operational way. In other words, he applied general concepts found in mathematics to library classificatory work rather than specific mathematical symbols, algorithms, and methods.

---

[24] A very strong sense of Ranganathan's "discovery" of documentation can be found in his *Library Tour 1948* (1950).

[25] "Micro-documents" was Ranganathan's name for parts of documents of any size or for relatively brief documents such as short scientific reports. "Micro-thought" was his way of denoting the topical content of "micro-documents" no matter how such items appeared. Focusing on his "discovery" of these concerns in documentation at this time in his life does not mean that he had not previously considered such things at all. There are ample indications in his works on cataloging and classification from the 1930s that show that he had considered such things at least theoretically. But it would appear that the importance of such matters and their impact on his classificatory work in an operational way did not occur until the late 1940s. Subsequent works such as *Classification and Communication* (1951a), *Philosophy of Library Classification* (1951b), and the two later editions of his *Prolegomena* (1957 and 1967) provide this broader sense of his classification mission.

Two particular aspects of mathematical ideas are especially significant in Ranganathan's approach to library classification. The first was his use of the axiomatic method, the second his use of ideas about numbers and infinity. The axiomatic method involves studying a topic or solving a problem by applying to it a set of axioms or self-evident principles. In mathematics and logic, the axiomatic method is considered basic to formal approaches to problem-solving, and in that context an axiom can be viewed as "a proposition that is assumed without proof for the sake of studying the consequences that follow from it."[26] When applied less formally, the axiomatic method refers to any approach to problem-solving that first identifies basic principles and then applies these principles rigorously to the problem at hand. It presupposes, of course, that one can identify the basic principles and that these principles are sufficiently fundamental for the purpose at hand.

Ranganathan studied mathematics during a period in which the axiomatic method was highly touted by the school of formalist mathematics, the dominant school of the era. Formalists concluded that an axiomatic approach to mathematics was not only critical, but also that it held some promise for solving all mathematical problems in a systematic fashion. It was in this light that David Hilbert (1862–1943)—one of the most prominent formalist mathematicians of the time and the proponent of a grand program for the solution of basic mathematical problems—attempted to reduce all branches of mathematical thought to fundamental axioms and postulates.

The axiomatic method appears to have found a ready adherent in Ranganathan when he took up mathematics as a career.[27] When Ranganathan switched from mathematics to librarianship, he found a weak form of the axiomatic method present in the classificatory work of Sayers (i.e., Sayers' indentification of "canons" of classification). Although this doubtless resonated well with Ranganathan, his experience with the axiomatic method as a much more critical issue led him to apply it to librarianship in general and to library classification in particular in a much broader and more thoroughgoing manner than Sayers.[28]

---

[26] This definition is taken from Random House (1987), where it is specifically listed as basic to mathematics and logic.

[27] What more could a young mathematician such as Ranganathan aspire to than to follow in the footsteps of so forceful a professional leader as Hilbert, the chief proponent of formalism? It is in this light that there can be little mystery as to why Ranganathan invoked Hilbert's name throughout the remainder of his life as one of the great leaders in that field.

[28] Near the beginning of his library career, Ranganathan identified and explicated the five basic laws of library science that in his view formed a basis upon which all other library operations and ideas could be built. He wrote of these in his *Five Laws of Library Science (1931)*. The laws were: "1. Books are for use. 2. Every reader his book. 3. Every book its reader. 4. Save the time of the reader. 5. A library is a growing organism." Ranganathan worked on the same basis in library classification where he identified eleven basic "laws" (i.e., "normative principles," including the five "laws" of library science), forty-three "canons," thirteen "postulates," twenty-two less extensive principles, and nine basic operations (i.e., "devices") and then referred to them constantly throughout his classificatory works. The most complete systematic representation of his classification laws and principles may be found in his *Prolegomena* (1967).

The principles (i.e., axioms) with which Ranganathan began were not specific mathematical formulae or algorithms, nor did they contain the attributes of mathematical entities. Rather they consisted of assertions about the characteristics of and relationships among books, libraries, and book users. With respect to library classification, they focused on specific operations and elements of the book classification process. In sum, Ranganathan adopted *the idea* of an axiomatic method for classificatory work, not the specific axioms of mathematics. He then applied that more general idea in a thoroughgoing manner to library classificatory entities, just as a mathematician would have applied axioms suitable to mathematical entities in a mathematical context.

A second and even more specific aspect of mathematics that Ranganathan applied to library classification was the use of the idea of numbers and infinity as an analogy for subjects and the universe of subjects.[29] The specific source of this analogy was the work of Georg Cantor (1845–1918), another mathematician cited by Ranganathan. Cantor, who was most closely associated with Ranganathan's earlier mathematical work, is generally credited with developing the idea of transfinite numbers. It was Cantor's work which, when combined with Hilbert's and others after the turn of the century, led to the field of axiomatic set theory. Cantor challenged long-standing ideas about the idea of infinity, especially the assumption that infinity was a singular and opaque notion that could be spoken of in only one way—that is, as a single sequence of points or numbers which simply had no end. Cantor argued that infinity could be viewed abstractly and could be manipulated in the form of symbols just like any other abstract entity. In reality, one could invoke the idea of multiple infinities of numbers (a conclusion initially derided by some in the mathematical community) and perform such operations on them such as comparing their respective size or membership (i.e., correspondence issues).

Ranganathan's more specific use of the idea of numbers and of infinity that is based on Cantor's work can usefully be divided into two parts. First, numbers, like points on a line, points in a plane, or points in a given space, are infinitely numerous and exist not simply as a single set of entities but as sets of entities in multiple infinities of points. Stated differently, an infinite number of points or numbers exists in multidimensional space, where each dimension represents a separate infinity of points.

Ranganathan used the same sense of points or numbers in multidimensional space as an analogy for the universe of subjects. The universe of subjects consists not merely of a single infinite structure of subject entities (an assumption which without reflection one might conclude is implied by the phrase "*the* universe of knowledge"), but rather of multiple infinities of subjects. In short, subjects, like

---

[29] Ranganathan almost always used the phrase "universe of subjects" in contrast to the phrase used here, "universe of knowledge." Upon reflection, it will be seen that "universe of subjects" is fully in keeping with Ranganathan's approach to the idea of infinities as described in the paragraphs to follow. Basically, because he did not see infinity as a singular entity, he could not refer to an infinity of subjects in the singular and holistic way that the phrase "universe of knowledge" appears to do. His eye was always upon the elements of that universe and, for him, the elements were the subjects that made up the whole.

points, exist in a multidimensional realm, where each dimension represents a different family or set of subjects. Ranganathan referred to such infinite sets of subjects by the term *facets,* and subsequently divided the facets into two parts: one part was a facet of basic subjects, the other part was an infinite number of facets of qualifications of basic subjects which he called *isolate* families.

Ranganathan's assertion that the total number of facets was theoretically infinite might seem to make the process of characterizing them impossible, but he also asserted that although their numbers are infinite, all such facets of subject isolates could be viewed as expressions of five basic kinds of facets—that is, in terms of attributes which he named energy, matter, space, time, and personality. The chief result of this approach to subject facets was to reduce their potential infinite number to a comprehensible and manageable number.[30]

The second part of Ranganathan's use of numbers and infinity had to do with correspondence relationships among infinities of numbers or points. The points in infinite sets can be compared in terms of their correspondence, and also can be transformed (i.e., mapped) from one infinity or dimension to another. Thus, the infinite set of points in a plane can be transformed or mapped to their respective positions on the surface of a sphere, or vice versa, and the infinite set of points on the surface of a sphere can likewise be transformed or mapped to their respective positions in a single linear sequence on a line. Ranganathan was most interested in the prospect of mapping from one dimension to the points on a line because of his awareness that the points on a line can be viewed as a model of the linear sequence of symbols in a classification system notation. The character of the numbers in such a system was especially important in that they typically consisted of a sequence of ordinal numbers (either integers or decimal fractions) which denoted position rather than quantity.[31]

Ranganathan applied the idea of transformation or mapping to subjects which exist in different dimensions or facets. If one assumes the subjects of the universe of subjects actually exist in the form of separate dimensions or facets and, more particularly, that they exist in the form of five basic kinds of facets plus basic subjects, then one can also potentially transform or map them from their respective separate facets or dimensions to some other dimension.

---

[30] The energy facet of subject isolates consisted of those aspects of basic subjects which had to do with the manifestation of action, the matter facet with the manifestation of the material or properties of basic subjects, the space facet with the manifestation of geographical position, the time facet with the manifestation of units or periods of time, and the personality facet with something akin to the essence of the basic subject but which operationally could be considered what remains when the other facets are first defined in any given instance.

[31] Cardinal numbers denote quantity so that the number "4" refers to four things, the number "10," to ten things, and so on. However, when numbers are ordinals, they represent not quantity but rather position within any one set of such points. Thus, the number "4" as an ordinal represents the 4th point, the number "10," the tenth point, and so on, in a set of such points. If the ordinal numbers are decimal fractions, then an additional feature is that new points may be interpolated between existing points in an infinite way. For example, given two points symbolized by .4 and .5, one could place still another point between them at .45 (or at any decimal fraction in the formula .4n) and still preserve the order of the original points.

More precisely, one can map these subjects from their respective facet positions to positions in the linear sequence of ordinal numbers which make up the notation of a subject classification system.

Two problems arise in this kind of transformation of subjects, however. First, the subjects within a single facet must have an order determined for them. Second, an order must also be determined for the facets themselves once mapped to a notation. The latter of these two problems was particularly intense because even the limited number of facet types that Ranganathan elicited could have potentially large numbers of different arrangements. Ranganathan solved the first problem, the order of subjects within a single facet, by establishing a research agenda (later a research facility) that over the years carefully analyzed the isolate families of subjects associated with specific fields of knowledge, enumerating them in their respective domains. He solved the second problem, facet order, by concluding that since all such facets were of five basic kinds, a "best" order could be chosen for them among all potential orders. He established a "best" order, which is known familiarly as his PMEST formula. Acording to this formula, when a basic subject with its isolate qualifications is expressed, values for the personality facet were to come first, values for the matter facet were to come second, those for the energy facet third, those for the space facet fourth, and those for the time facet last.

Ranganathan's analogical use of the characteristics of numbers and infinity enabled him to analyze the universe of subjects and then portray this universe with the kind of complexity and sophistication that had been increasingly attributed to subjects during the first half of the twentieth century. Further, his faceted structuring of classification notation, called notational synthesis, allowed him to incorporate that complexity into the classification process. It also supplied the latitude necessary for accommodating a host of new areas of knowledge, areas that were being sought in documents, especially in scientific literature. He called this process an analytical-synthetic approach to classification, thus capturing the two basic aspects of his work. Much more could be said about the entire range of his methods and techniques, of course, but that is beyond the scope of the present treatment. What is important to note is that Ranganathan's analytical-synthetic approach to subjects and to the universe of subjects, regardless of its origins in a mathematical analogy, proved to be a watershed for the classification work that came after him.

Given Ranganathan's extensive contribution to library classification theory and technique, it might seem superfluous in the next breath to offer criticisms of his work. But the fact is that Ranganathan's approach to library classification theory and technique also had negative aspects that should be mentioned. First, in the process of using his analogy for the universe of subjects, Ranganathan also helped to promote what generally may be called the "atomization" of subjects that has taken place since World War II. Atomization of subjects refers to focusing on subjects individually apart from the subject structures in which they might be found, regardless of whether those structures are thought to be portrayals of the universe of knowledge or are merely the structure of the complex content of some individual document.

The origin of this phenomenon ultimately goes back to the change in how subjects are viewed that arose in the documentation movement. The growing expectations of both information workers and information users to retrieve information-bearing entities, not merely as the repositories of main or dominant subjects, but also as the repositories of numerous subjects of all kinds—some extensive, others not, some dominant, others minor, some with simple accepted names, others named only by resorting to complex constructions of simple names—has led to an enormous increase in the recognized population of the universe of subjects. This has led, in turn, to the necessity of expending enormous resources to use classificatory subject structures to bring order to the world of information as a whole. In the face of such requirements, and with the adoption of computer technology and its capacity to search for individual terms, an alternative has arisen since World War II which allows us to deal chiefly with individual subjects without considering the classificatory structures of which they are elements—hence their "atomization" into their individual parts and units.

Ranganathan never promoted treating subjects apart from classificatory subject structures because he was ultimately committed not only to the unity of the universe of subjects but to the idea that the subject content of a given document was itself a unity.[32] However, by dismembering traditional subject structures through his approach to the universe of subjects, he also helped focus attention on the parts of the whole—that is, on subjects individually—quite apart from any kind of a structure, whether conceived in an older one-dimensional hierarchical way or in Ranganathan's more complex faceted structure. To say this is not to cast aspersions on his work. It is simply to admit that his approach to the universe of knowledge can be viewed as something of a two-edged sword. Buried in his multidimensional universe of subjects are the seeds of dismantling that same universe as a unified structure. Once dismantled, there are many who have concluded that reconstructing this universe is neither worth the effort nor necessary.

Second, neither Ranganathan himself nor those who have adopted his approach to subjects have ever critically analyzed the analogy at the base of his approach. They have never asked themselves whether it is an adequate way to represent subjects in the universe of knowledge. In this respect, one might indeed ask, How are subjects like points in multidimensional space? The analogy appears to fit when it emphasizes the fact that subjects tend to have close associations, such as groupings of time periods, places, objects, processes, and so on. But the analogy appears not to fit, or at least not to fit very well, once such groupings have been devised and the subject variations and exceptions in them cease, or once the very act of devising such groups tends to remove such variations and exceptions from them. But sub-

---

[32] Ranganathan's commitment to the unity of the idea of a subject in a document is in many respects a necessary component of his goal of deriving classification numbers that are "co-extensive" with the subject scope of the document. The only way that he was able to overcome that more holistic approach to the matter was to define parts of a document as documents in and of themselves—that is, as "microdocuments" containing "micro-thoughts." Even that resolution does not accommodate very well, if at all, the idea of a subject theme that appears and reappears in a scattered manner throughout a document but never in a concentrated place in one location in the document.

ject variations are a function of language and, especially, of semantics. It is this realm of language and meaning, in fact, that makes language so "ragged." However, it was precisely the raggedness of language that appears to have made Ranganathan uncomfortable—witness his joy at being able to assign ordinal numbers to subjects in a way that somehow "fixed" their meanings in time and space. In the end, there is strong indication that Ranganathan's use of a faceted structure of subjects may well have represented his need to find more order and regularity, in the realm of subjects, than actually exist.

Third, like nearly all library classificationists before him, Ranganathan vigorously pursued the goal of finding the *one best* subject classification system, the system which, once it was recognized for its excellence, would be adopted widely and used as a standard for all forms of subject retrieval of information. As a consequence, Ranganathan would denigrate, sometimes without a healthy respect for the historical development of classification, other contemporaneous systems, in favor of promoting his own system. He also paid insufficient attention to the strong points of other systems or to the weak points of his own. The chief problem with this approach is that it was predicated on the assumption that a single best system will reflect the way the universe of knowledge is actually arranged, that such a system is attainable, and that it is the best solution for all retrieval uses of classification. Any close scrutiny of these assumptions will reveal, however, that they are holdovers from the nineteenth-century beginnings of modern library classification, and, as such, at the very least, should be carefully examined.

The Legacy of the Library Classification Theorists

The work of four library classificationists has been briefly examined in order to provide some sense of the development of library classification theory and technique during the twentieth century. Their work, when combined with the rise of a new approach to the idea of a subject in the twentieth century, has provided the bedrock for library classification theory development since their time. One could go on from here, of course, to see how the specific ideas of each of these four theorists have influenced library classification development since their time, but that would be beyond the scope of the present treatment. What is appropriate, however, is to summarize the general impact of their work on subsequent library classification development. In this respect, their work can be viewed in terms of three significant themes: 1) characterizing the universe of knowledge as a basis for legitimizing library classification; 2) defining more narrowly the purpose of library classification; and 3) discovering an appropriate operational technique for library classification.

Characterizing the Universe of Knowledge as a Basis for Legitimizing Library Classification

When modern library classification began in the late nineteenth century it constituted an essentially pragmatic activity concerned primarily with books and other information-bearing entities. A library classification was a way of arranging (i.e., sorting) such materials in a library collection so that the library could serve the goals of the modern library movement. Now, even Cutter recognized that more was involved in library classification than simply sorting books. For example, he wrote that once having sorted books into classes, a classificationist must then go on to sort the *classes of books* into a structure of classes.[33] However, it is not until E. C. Richardson that one finds the first extended inquiry into the nature of that other set of objects—*classes of books*. Classes of books are really ideas or concepts found in books and other information-bearing entities serving as attributes upon which sorting is based. When focusing on the attributes alone, however, the "objects" a library classificationist is sorting have ceased being merely the physical objects in which such ideas or concepts are found, becoming instead the ideas or concepts themselves.[34]

Richardson concluded that if a large part of what library classificationists do is to sort ideas or concepts, then it behooves them to know as much as possible about those ideas and concepts, including their origins and to what they refer. He believed that the ideas or concepts featured in library classification refer to the "things" of actual existence, and that those same "things" of actual existence have their own natural order which library classificationists would be wise to follow—be that an evolutionary order, a historical order, or any other kind of order. Richardson concluded, in other words, that the proper study of library classification is the universe of knowledge itself.

This kind of thinking may not seem very important to us today because library classificationists have expounded on the universe of knowledge extensively since Richardson's time. Sayers concluded, for example, that no matter how the universe of knowledge is structured, that structure is ultimately strictly logical in character. Later, in the 1920s, Bliss expanded Richardson's idea of the relationship of classification categories to the "things" of actual existence into a grand theory of Nature, a corresponding grand theory of scientific consensus related to the investigation of the things of Nature, and a reasoned approach to arranging the categories which arose from that consensus—that is, scientific fields arranged by the idea of "gradation by specialty." Still later in the 1960s, members of the British Classification Research Group resorted to ideas of integrative levels spun off from Ludwig von Bertalanffy's General System Theory in order better to articulate Richardson's and Bliss's vision of the "things" of actual existence. At the turn of the century, however, this kind of thinking was not common. It took the explanations of these four writers to legitimize discussions of the universe of knowledge. The result of

---

[33] In his *Rules for a Printed Dictionary Catalogue* (1876b, p. 12), Cutter stated that the act of classifying books by subject entailed the following processes: "1. Bringing books together which treat of the same subject specifically. . . .2. Bringing books together which treat of similar subjects. . . . Bringing subjects together so as to form a class. . . . 3. Bringing classes together so as to form a system. . . . The three steps are then 1. Classifying the books to make subject-lists. 2. Classifying the subject-lists to make classes. 3. Classifying the classes to make a systematic catalogue."

[34] Classes of books are not limited solely to ideas or concepts, of course. They may consist of other attributes such as genre or publication formats. But at this stage in library classification development, the role of concepts or ideas as classes or objects was the most important concern.

their efforts was to help make the universe of knowledge an object of investigation in and of itself, a phenomenon worthy of investigation; one could produce hypotheses about it, speculate as to how to characterize change in it, and so on.

As the universe of knowledge has become a legitimate object of investigation, classification theorists have discovered that it is a much more complex phenomenon than was previously supposed. In fact, the closer Richardson, Bliss, Sayers, Ranganathan, and those who have followed them got to the phenomenon, the more complex it appeared and the more difficult it was to describe. Much of this new sense of complexity was due to the new approach to subjects that arose from the documentation movement and elsewhere, where subject complexity is a direct reflection of the way in which scientists, engineers, and specialists of all kinds search for information.

Two derivative results subsequently arose from this effort to characterize the universe of knowledge. One result has been an ongoing attempt to characterize the universe of knowledge in all of its complexity by means of some suitable analogy. Various analogies have been propounded in this respect, some of which have been fairly traditional. For example, Richardson, Bliss, and Ranganathan at one time or another resorted to the analogy of a growing tree as a way to speak of the universe of knowledge, the tree's various branches representing its divisions and the tree's system of branches representing its complexity. Since their writings, other analogies have been put forward, one of the most striking being that of E. J. Coates in his *Subject Catalogues* (1960, p. 32) where he pictures the universe of knowledge in terms of a geographical metaphor in order to show the growing complexity of the universe of knowledge since the nineteenth century.

The most striking analogy of all, however, is Ranganathan's mathematical analogy of a multidimensional realm of subjects, a matter already discussed at some length. Ranganathan fundamentally changed the way the universe of knowledge (i.e., Ranganathan's universe of subjects) is viewed in library classification theory and technique by shattering the previous view of it as a singular, one-dimensional hierarchical structure and replacing that view with one that is structured in a complex, modular, faceted way. Since Ranganathan's day no library classificationist has failed to view the universe of knowledge in this new complex arrangement of faceted structures. At the same time, it should also be noted that few, if any, library classificationists have adopted Ranganathan's picture of the universe of knowledge in terms of the mathematical analogy behind it. Library classificationists appear to have adopted Ranganathan's view chiefly because it offered a more satisfactory way to view and deal with subject complexity than was offered by any previous approach.[35]

---

[35] I for one am also tempted to conclude that Ranganathan's faceted universe of subjects has been adopted as much for what appears to be some sort of intuitive correctness as for any other reason, but a claim of this sort is little more than unsupported speculation. Still, one cannot easily miss how nicely his formulation of the facet idea seems to fit certain subject areas, and this may be reason enough to adopt it as a standard approach, regardless of whether it has any sound scientific basis or whether it always serves well.

The other derivative result of the effort to characterize the universe of knowledge has been to encourage a subtle shift in how the library classificatory enterprise views itself. This shift can be seen if one compares nineteenth-century library classification with post-1950s library classification. In the former, the classificationist's relationship to the sciences was primarily to view them as objects denoting areas of knowledge. In the more recent period, one finds library classification researchers not only connecting the complex phenomenon of the universe of knowledge to specialists as users of information, especially specialists as scientists, engineers, and other scholars, but also identifying themselves with those same users—that is, with scientists, engineers, and scholars. In the process they have come to describe their work not merely as the classification of the sciences, but also, and more specifically, as the act of scientific classification, with appropriate emphasis on the intellectual weightiness of the endeavor and the importance of correct "scientific" methodology.

Defining More Narrowly the Purpose of Library Classification

The second theme that arose from the development of twentieth-century library classification theory was the attempt to define more narrowly the purpose of library classification. When modern library classification began in the nineteenth century, its purpose was intertwined with the educational and cultural objectives of the modern library movement. That movement took as its primary aim the cultivation of people's minds so as to produce citizens for an enlightened democracy. Within that ideological context, library classification had more than one role. On the one hand, it served as a tool of education, instructing patrons over a period of time as to how knowledge was structured. This educational goal was viewed as important because of the belief that all proper subject access to materials was classificatory at its very heart. Thus, where patrons had access to printed catalogs and those catalogs were classified, they would, when reading such catalogs, also receive instruction about the idea of the universe of knowledge and its structure. On the other hand, library classification also provided a useful retrieval tool to identify the best books by the best authors in each of the several departments and branches of knowledge. It served, in short, as a relatively undetailed but nonetheless effective map for locating the best reading for the process of the intellectual cultivation of library users.[36]

In the twentieth century, documentalists became aware of the new idea of subjects and the corresponding complexity of subject structure in the universe of knowledge. This awareness, when combined with their interest in using classifica-

---

[36] The educational goal of classification in catalogs is discussed in Miksa (1983a). As to the retrieval goal, it appears likely that the original target audience for library shelf classification systems was the librarian himself or herself and not the end-user who is now ordinarily thought of as the target. Prior to the 1890s library patrons did not ordinarily go to the shelves. Instead, librarians picked out books for them and a classification scheme was a useful map for quickly getting to the appropriate section of the collection. It was only after open shelves came to be standard after the turn of the century that library classification systems became thought of as tools for end-users.

tion as a way to provide information needed by scientists, engineers, and other specialists, forced them to seek ways to define more narrowly the purpose of library classification. As the twentieth century has proceeded, this more narrowly defined purpose has become document retrieval in and of itself, not document retrieval intermixed with other goals. This shift occurred because of the new urgency for retrieval—that is, scientists, engineers, and specialist scholars of all kinds needed help to pursue their work, make contributions to society, and take the next steps in the pursuit of their activities. In short, the purpose of library classification was to facilitate what Fritz Machlup would later characterize as the instrumental or practical use of information; it was no longer simply a tool to serve more general cultural or intellectual approaches to information.[37]

The emphasis on document retrieval and, more specifically, on document retrieval for the instrumental use of information, has been bolstered by two corresponding beliefs. They are called beliefs because no one has ever really tested their validity. The first such belief is that end-users naturally search for documents within a hierarchical classificatory context (i.e., up and down or across chains of topics). The second belief is that the survival of civilization as we know it is wedded to humankind's ability to organize knowledge and the documents which contain knowledge for the instrumental use of the information they contain, and that such organization requires systematic, hierarchical structuring of the universe of knowledge. Two of the most eloquent statements of the latter belief can be found in the beginning sections of Bliss's *Classification of Knowledge and the System of the Sciences* and in S. R. Ranganathan's *Classification and Communication.*

Discovering an Appropriate Operational Technique for Library Classification

The third important theme to come out of the development of twentieth-century library classification theory was the discovery of an appropriate operational technique for library classification. Here, the contribution of Ranganathan is plainly supreme, for it was his blending of a multidimensional universe of knowledge with faceting as a technique of notational synthesis that has provided twentieth-century theoretical library classification with its most powerful new tool.

I could continue at length to chart the course of developments such as the foregoing, but that would occupy the content of still other whole volumes. What is most important here is to assess the impact of these various kinds of developments on the system that is at the center of consideration here, the Dewey Decimal Classification.

---

[37] Machlup put forth his categories of knowledge (i.e., of information use) in the first volume of the revised version of his massive work on the production and distribution of knowledge in society (Machlup 1980, pp. 108–9). An explication of his categories which interprets his instrumental knowledge as information that allows one to "take the next step" in some activity can be found in Miksa (1985, p. 162).

The DDC and the Development of
Library Classification Theory

Tracking the relationship between the DDC and the development of twentieth-century library classification theory is fairly straightforward. Before the 1950s the relationship was indirect, whereas since the 1950s the relationship has become increasingly direct. The reasons for this difference reside in the general historical context in which the DDC developed.

The DDC and the Development of Library Classification
Theory before 1950

From its beginnings to the 1950s, the DDC was essentially an expression of the general modern library movement as it developed in the United States in the late nineteenth century. This means that its usefulness and justification were seen primarily as a way to serve that more general library movement. Some mention of this has already been made in the section above where I discuss the rise of a new view of subjects. For the same reasons that the new view of subjects took a long time to become a forceful idea in the general library movement, so also did it take a long time for the same view of subjects and developing library classification theory to affect the DDC. The general library movement was at its core an educational venture. It had to do primarily with the ideal of helping create an enlightened citizenry. Its primary focus was, in short, on what knowledge in books *did to people,* not on what people using the books *could do with* knowledge. In terms of library classification goals, that meant that the DDC's principal role was to be a general mapping system for library collections so they could in turn fulfill their educational goal. The DDC was not supposed to function as a precise retrieval device for a special subset of users made up of scientists and other specialists, nor was it supposed to provide a detailed mapping of the universe of knowledge for very large library collections.

The DDC was affected by this orientation for as long as those who controlled the system were focused on the ideals of the general library movement and saw the role of the DDC in terms of service to that movement. Melvil Dewey as the founder of the system and others who worked closely with him kept the system true to that purpose well into the 1930s. But even those who followed that original group during the 1940s and into the early 1950s saw the mission of the system in much the same way.

The only impact that the growth of library classification theory and technique had in the foregoing context was to exert what I earlier called "pressure" on the DDC to accommodate specialized fields and larger library collections (such as those in the major public libraries and universities) by creating more detailed classification schedules. Responses to the pressure were essentially pragmatic, however. They were made without any real sense of the larger issues at stake and, accordingly, the results seemed haphazard. Thus, by the 13th edition, the level of specification in the system had become very inconsistent, the collocation of subjects had become more and more dated, and little attention was paid to the new classificatory techniques of the kind found in the UDC, even though these techniques held some promise for dealing with the complexity that ensued when the hierarchical structure of the system was deepened. The 14th edition attempted to correct the issue of inconsistent specification but, in the process, expanded the system enormously without any correction of collocation problems and without any recourse to techniques which would have helped handle the complexity that arose from deep hierarchical structures. The 15th edition, although correcting issues of dated collocation and updating terminology, confused the situation even more because of its troubled sense of goals for a standard edition.

In many respects the foregoing developments are understandable given the larger historical framework in which those who were responsible for library classification theory development and those who were responsible for the DDC found themselves. To begin with, classification theory in the library field remained relatively undeveloped throughout this earlier period. Only a relatively small coterie of librarians discussed library classification theory. Some of those who were involved in the actual creation and maintenance of specific library classification schemes discussed various techniques that might be used in library classification systems. However, these discussions most often took place as comparisons of schemes in the ongoing competition to find the one best library classification system and not as investigations of the theoretical underpinnings of such systems.

The lack of widespread theoretical discussion should not seem strange, however. In reality, before the 1930s only a few scattered voices were raised about theoretical issues of any kind in librarianship and even these voices did not normally express their ideas in the context of controlled research methodologies. This situation began to change somewhat during the late 1920s in the United States with the founding of the University of Chicago Graduate Library School, but that was only after twelve editions of the DDC had been issued. It was not until the establishment of more library school doctoral programs in the late 1940s and 1950s (e.g., the Universities of Illinois and Michigan in 1948; Columbia University, 1952; University of California at Berkeley, 1955; Western Reserve University, 1956; Rutgers University, 1960) that general theoretical discussions began to infuse the field. Within this general context, the fact that theoretical discussions of library classification appeared only at the end of the 1940s and the early 1950s is much more understandable. Their appearance at this time fit quite naturally into the broader general development of the theoretical side of the field.

In a parallel way, and despite the concerns of those in the documentation movement and in special libraries, the status of science, technology, and other scholarly work, and the research that is ordinarily associated with those areas, also remained relatively undeveloped through the first three decades of the twentieth century. Ultimately, however, these areas were the driving force behind the rise of library classification theory. Providing information service to the persons involved in such research caused the rise of the documentation movement in the first place. However, without the corresponding social esteem and governmental support which promoted science, technology, and scholarly research of all kinds, there was little reason to view the needs of these areas as critical to the library movement to which the DDC owed its rationale, let alone to make adjustments in the DDC so that it could serve such needs.[38]

Ultimately, the information needs of scientists and specialists, the realm which prompted library classification theory discussion, remained undiscovered by librarians in general until after World War II. And that discovery paralleled the rise in the social role and importance of science, technology, and widespread scholarly research in American society. That societal recognition occurred because of the information needs prompted by World War II and the Cold War which followed it.[39] It was at this point that S. R. Ranganathan, the last of the four theorists to be discussed here, was able to combine successfully the new approach to the idea of a subject which had been percolating for half a century with other seminal ideas of his own in such a way as to bring library classification theory into a position from which it could develop on its own.

Both of these historical developments, the occurrence of widespread discussion of library classification theory and the recognition by the general library realm of the information needs of scientists and other scholarly specialists, occurred in the 1950s. The correlation of these developments with the life of the DDC is also striking, for within a decade of their appearance the DDC was already well on its way into a new period of change driven by library classification theory and technique development.

The DDC and the Development of Library Classification Theory since 1950

Since the 1950s the impact of library classification theory and technique development on the DDC has been very direct. The most obvious results, discussed in detail at the end of the first section of this paper, have included enormous and growing efforts to keep the scheme up-to-date in its terminology and subject collocation, and the increasing use of subject faceting and notational synthesis in the system. The impetus to adopt such changes arose in large part because of the prominent role that the Library of Congress assumed in the maintenance of the system and, especially, because of the work of Benjamin Custer who almost single-handedly set the system on its present course. There are many critics who have complained that change in the system has not been either fast enough or thorough

---

[38] Some accounting of the general rise of research and its impact on research libraries can be found in Miksa (1987). An accounting of this phenomenon within the library field in general in the twentieth century has yet to be written, however.

[39] The shift that took place which resulted in this social recognition of science and technology in society has been investigated from more than one standpoint. Some have focused on national science policy, for example, and the important role played by various persons and agencies, not the least among them being the work of Vannever Bush and the creation of the National Academy of Science. Agencies closer to the field of librarianship include the Science and Technology Division of the Library of Congress, the seminal work of the American Documentation Institute, and within the latter the work of Mortimer Taube, C. Dake Gull, and others.

enough, of course. But, the steady rather than headlong pace of change in the system has been controlled by a balance perceived as necessary between serving the needs of the more general educational side of the library movement and the libraries which primarily adhere to that role and serving the needs of document retrieval for specialists and scholars of all kinds.

Assumptions That Remain

It might seem that, as the DDC has developed a base of theory and technique since the middle of the twentieth century, it is now much more in the mainstream of twentieth-century library classification development. However, one must be cautious about assuming that this picture is a totally happy one. Despite all of the advances that modern library classification has now encompassed, it includes three assumptions that remain unquestioned and untested. These assumptions could be detrimental to the work of classification in general and to the future of the DDC.

First, for the most part, twentieth-century library classification has assumed that knowledge categories are by nature hierarchical and logical in a classical, systematic sense. This belief arose from assumptions that the modern library classification movement adopted in the late nineteenth century from the movement to classifiy knowledge and the sciences. These same assumptions have been followed for most of the twentieth century as well. It is this assumption, surely, which causes librarians all too quickly to equate classification with systematic hierarchical systems like the DDC and to exclude other systems such as subject heading lists or cluster production systems even though such systems also qualify as classification systems. It strikes me, however, that enough is now known about classification in general and about human mental categorization practices in particular to conclude that logical, hierarchical structures are only one way to structure knowledge categories in relationship to one another.

Second, for the most part twentieth-century library classification has subscribed to the idea that somewhere, somehow, we can, or should try to, produce the one best classification system that will serve all purposes. This conclusion in turn rests on the assumption that one's attempt to classify knowledge is really the act of discovering classificatory structures that are inherent in nature or that exist in society in some permanent way. The fact is, however, that knowledge organization systems appear to be inherently personal, and while many people in groups might agree in great part on how such systems should be structured, there is in the end no one best classification of knowledge system in any absolute sense. There are only alternative arrangements that serve one or another purpose or person for better or for worse.

Third, because twentieth-century library classification has subscribed for so long to the idea that the purpose of library classification is document retrieval and, more specifically, document retrieval of the most exacting sort in order to meet the needs of the instrumental use of information, it has been difficult at best to see all of the other many legitimate possibilities for which classification can be used. The classification of any set of entities, regardless of whether they are documents, concepts or objects of any kind, ultimately yields a system of categories that serves at its most fundamental level as a reference system for those objects. The storage and retrieval of objects constitutes only one of those uses, however. Other purposes include the use of a reference system as an aid to memory and teaching, as a basis for communication, and as a way to develop hypotheses about the origins, nature, and relationships among the objects classified, to name only a few. In the earliest decades of modern library classification development, no small part of its use had to do simply with the education of people who used such systems, a matter already noted. Here, it is enough to suggest that some recovery of the broad range of purposes to which library classification might be put would be in order.

The DDC has been affected over the past six editions (i.e., since the late 1950s) by modern classification ideas and concerns, which are based on the three assumptions just discussed: knowledge categories are by nature heirarchical and logical; one best classification system is achievable; and document retrieval is the main purpose of library classification. I conclude, therefore, that those who are responsible for the DDC do an injustice to the system in adhering to these assumptions without serious reflection. They would better serve library classification in general and the Dewey Decimal Classification in particular if they were to adopt a much more open and questioning stance toward these assumptions when considering the future of the system.

PART IV

Summary and Conclusions:
The DDC and the Post-Modern Library

I began this account by discussing the development of the DDC in terms of three distinct periods in its growth—Editions 1 to 6 (1876–1899), the period in which the system was constituted; Editions 7 to 15 (1911–1953), a period of conflict and inconsistent change; and Editions 16 to the present (since 1958), the period during which the system has made enormous advances based on developments in twentieth-century library classification theory and technique. I then went on to describe a broader context for the development of the system in order to understand better why Melvil Dewey, the DDC's author, had created this kind of a system and not some other. The answer required an exploration of the idea of the "universe of knowledge," which, in turn, required an examination of the relationship of the DDC and other early modern library classification efforts to the nineteenth-century movement to classify knowledge and the sciences. Finally, I examined the relationship of the DDC to library classification theory and technique, as that theory and technique developed in the twentieth century independently from the nineteenth-century movement to classify knowledge.

We may conclude in summary that Melvil Dewey created the DDC in the form that he did because that is how knowledge was viewed in the times in which he lived, that is, as a relatively simple, one-dimensional hierarchical structure of subject categories with the most general subjects at the top of the structure and the most concrete categories at the lowest levels of the structure. We may also conclude that Melvil Dewey (as well as others who worked on library classification well into the twentieth century), was not a theoretician of the universe of knowledge. Rather he was more the practical library classificationist and adroit businessman who invented a highly successful tool that could be used with great efficiency to promote the work of the modern library as it evolved in the late nineteenth century. For the first seventy-five years of its existence, that is, through the publication of the 15th edition in 1951, Melvil Dewey's practicality and business sense ultimately put the development of the DDC as a system on a course that was relatively cohesive, although not without conflict.

What Melvil Dewey did not appear to understand (and he was not alone in this) were the implications in the increase of scientific and scholarly publishing and the rise of a new view of the idea of subjects which appeared after the turn of the century. He also failed to understand (and again he was not alone) how these developments would affect the urgent need to supply an efficient information service for an emerging world of scientists, technical workers, and scholarly specialists of all kinds. Nor did Dewey really comprehend the implications of the rise of library classification theory in its own right as the twentieth century progressed. The new view of the idea of subjects, especially in its relationship to information service for the scholarly establishment, was subsequently merged in the late 1940s and 1950s with the development of library classification theory in the work of S. R. Ranganathan and those who were inspired by him. That convergence has been the basis for changes taking place in the DDC since that time.

Given this course of events related to the history of the DDC, we might be tempted to conclude that the future of the DDC is assured, at least to the extent that its development stays on its present course. But there are good reasons to question this conclusion, reasons which are suggested by the appearance of what may be called the "post-modern" library.

The Post-Modern Library

The "post-modern" library has two separate but intertwined meanings, the first having to do with changes in libraries due to the growing potential of new information technologies, the second having to do with implications for library classification development due to a particular viewpoint of the present age.

With respect to the first meaning, it now seems evident that the library which we all know—the library which has been evolving for over a century since its beginnings in the late nineteenth century, which was a direct response to the rise of the modern print publishing culture and reading markets, and which has long been called the "modern library" as a way to distinguish it from the expression of the idea of the library which preceded it—is now going through a period of dramatic reformulation. This reformulation, begun just after World War II, has been accelerated with the widespread introduction and use of computer and telecommunications information technology in society. This acceleration has increased even more rapidly since the rise of the World Wide Web in the present decade. The new technology has the potential of creating a new expression for the idea of the library, one which for the first time since the early nineteenth century can function in terms of what is conveniently characterized as a "personal-space" rather than a "public-space" information environment.[1]

The library as a public-space information environment is what especially characterized the modern library created at the end of the nineteenth century. That form of the library, the library as social institution, gave the citizens of society both the right and ability to access information. The library as social institution meant that individuals did not have to spend enormous amounts of money to assemble extensive collections of printed information resources. It placed the responsibility for the cost of such collections and suitable access mechanisms primarily in the public realm. Also, the public space library could create the administrative structure needed to achieve reasonable and responsible efficiency in its operation.

The modern library was a marvelous invention for its time. But it also necessarily incorporated into its operation a layer of intractable complexity. In order to make such a public-space information access environment viable, the chief

---

[1] The comments in this paragraph and elsewhere here on an emergent kind of electronic library are based on Miksa (1996).

target for the library's operations had to be heterogeneous rather than cohesive groups of information users. In short, information resources collections and information access mechanisms had to be based on amorphously defined groups of information users rather than on individual information users or on small cohesive groups of information users. If the latter had been the case, needs could have been closely defined, and collections and mechanisms could have been tailored to the special needs of individuals or small groups, thus enabling the library's operations to function with great precision.[2]

In contrast to the modern library, the now emerging electronic library is made possible because of the growing presence of a networked world of information resources and a technological infrastructure that permits access to that world of information from one's own computer. In this new realm of electronic rather than print information, the selection of information resources and the mechanisms which make such resources so easy to access and use, has the potential of being tailored to the individual or individuals for whom such a library is especially created. Hence, it is fitting to view this form of the library as a personal-space information access environment.

This new form of the library will probably not make the public-space library of the modern period disappear, at least not in the near future, but it will make it possible for individuals to have their own personal libraries of whatever network-available information they wish to include. The access mechanisms to that information will suit them individually without concern that such access mechanisms might be unsuitable for someone else. One way to characterize the impact of this new emerging form of the library is to speak of it in terms of two additional laws which might be added to Ranganathan's own thoughtful laws of library science—"Every person his or her own library" and "Every library its own user."

The Post-Modern Age

The second meaning of the term "post-modern" reflects a theme in contemporary thought which emphasizes that a historical shift now appears to be occurring in society, a shift which will eventually place the library of the future in an entirely different historical and philosophical context. That shift is generally described as the movement of society from the world view of the modern period in history, which is ordinarily dated from the Renaissance and European Enlightenment, to a new world view denoting a new period in history, which is sometimes simply called the "post-modern" age.[3]

---

[2] In some respects, the rise of the special library after the turn of the century and certainly its development after World War II can be viewed as an attempt to control this situation by narrowing the target user group of a library and its information access mechanisms.

[3] The following account of post-modern thought is necessarily brief and will probably offend anyone who has read extensively on the subject. The fact is that a wide variety of positions exist as to what the idea means and how it may be applied. The account here represents my own attempt to summarize the topic in a way relevant to the present discussion, hopefully without doing the subject too much injustice.

One striking characteristic that distinguishes this newly emerging age from the past is what appears to be a slow but insistent questioning of, and in some instances even turning away from, the epistemological point of view basic to the modern period of history. That epistemological point of view has to do primarily with assumptions as to how people view and arrive at the "truth" about the world and humankind. In the modern period, an assumption has long been held that the world and humankind exist independently of any human observer and that, given this assumption, observations of the world and of humankind can be made in an objective manner. The result is the belief that we have the potential to discover truths about the world and humankind that are objectively accurate. Within this viewpoint, science has been considered the highest form of such truth-seeking, because it has always been assumed that the entire process of scientific investigation, observation, inference, and conclusion can be controlled, ideally at least, in a manner sufficient to safeguard against grievous and uncorrectable error.

In contrast to this viewpoint, post-modern thought generally concludes that the world and human relationships do not exist independently of an observer. Instead, the search for truths about the world and about humankind is always colored by the participation of the observer within the realm being observed. And further, arriving at truths about the world and about humankind is actually an involved process of human discourse which is, in turn, subject to various human propensities, not the least of which is the need to exercise power over the world and over one another. Given this context, assertions about the truth of some matter or another and especially about the truth of matters regarding humankind are fundamentally relative, with a propensity to serve human convenience. In short, truth is not absolute in any sense and society must reformulate itself to accommodate this factor.

The implications of this epistemological characteristic of a post-modern age are especially profound for the classification of knowledge, because the latter is in its own right an elaborate assertion about the collective knowledge of humankind. First, the post-modern point of view rejects the idea that there is some rendition of the universe of knowledge that reflects some absolute structure of subjects and their relationships, or that mirrors some absolutely accurate reference to what Richardson calls the "things of existence." This post-modern point of view would also reject any claim of absolute referential truth for any other assertion about the world or humankind, concluding instead that all arrangements of the subjects of the universe of knowledge are better spoken of as conveniences that arise from human discourse at given times, and that they vary among themselves with all the differences that arise from individuality in society. There is consequently no *one best* classification of knowledge system—that is, best in the sense of being accurate in any absolute sense.

Second, since the process of human discourse (by which such renditions of the universe of knowledge are made) is deeply affected by human propensities, this point of view also suggests that knowledge classifications, which are attempts to categorize the "things of existence," also show evidence of such propensities. In short, there can be no absolute conclusions as to what constitutes the best choice and placement of subject classes in such a structure or even about what constitutes a legitimate area of knowledge such as a discipline or a subdiscipline. Instead, there can only be arrangements and structures that are of convenience to those who make them. This accords with the notion that classificatory structures in general and knowledge classification systems in particular are not so much exercises in ascertaining what is in the world as they are exercises in "losing information" so as better to build a construction of reality that is satisfying for the moment or for a given set of purposes.

The Implications of Post-Modern Ideas for the DDC

These two meanings of post-modern—the creation of the personal space library and the belief that everything is relative—have profound implications for the future of the DDC, even if they contain only a grain of validity. One obvious implication is that in a post-modern electronic library of the kind described here, the collocation and retrieval functions of library classification in general and of the DDC in particular will change. There will be much less need to categorize large collections of primarily print information sources for an amorphously defined general information user in a public-space information access environment. Rather, the primary functions of library classification systems will be to categorize large and growing collections of (i.e., links to) electronic information sources for individual information users (or for small cohesive groups of information users) in personal-space information access environments. How will the DDC relate to this kind of a library? For example, can the DDC be fashioned as a product for end-users and their personal-space libraries? Can it be fashioned to accommodate widely variant individual use? One can imagine a scenario where a few highly developed sections of the DDC might need to be made available on an individual basis, while other sections remain underdeveloped and still other barely extant. Or, perhaps the DDC may need to be made available with highly supple mechanisms that would allow an individual without training in library classification to choose and arrange the categories of the DDC, displaying their relationships and shaping their levels of specification to his or her own liking.

Another even more profound implication for the future of the DDC arises from the epistemological issues raised by the idea of a post-modern age. If there is no *one best* approach to take when organizing the subjects of the universe of knowledge, what should be the role of a system like the DDC (or for that matter any other modern library classification or knowledge classification system) with respect to establishing a distinct order and structure of subjects? This role has been paramount in the history of the DDC, which is, after all, a system that supplies libraries with classification copy so they might more efficiently organize their collections. This question can be answered in two parts, one having to with the creation of various levels of specification in the system, and the other having to do with alternative arrangements, including the choice and collocation of categories in the system.

Specification Levels in the DDC of the Future

First, even if ultimately there may not be *one best* system to classify knowledge, a basic or "standard" arrangement of the DDC can still be devised for use by those libraries and individuals who find such a standard arrangement suitable for their own purposes. One complicating factor, however, would be the necessity to consider once again the role of a standard edition of the DDC, especially with respect to levels of specification. Here, valuable lessons can be learned from the discussions of the 1920s surrounding the call for creating three basic editions of the DDC, and from the experiences which arose from creating the 14th and 15th editions of the system. Instead of struggling with choosing some artificial level of specification, I believe that a standard edition would be better defined as the maximum level of specification possible in the system, combined with a suitable device to allow for a variety of lesser levels of specification so that libraries and individuals can use the system at whatever level of specification they wish. When the DDC existed only in print format, a single choice was a necessity. However, having the DDC in electronic form means that a mechanism can be created in the system that would allow it to be displayed and used at whatever level of specification a user desired. A mechanism of this kind in the electronic version of the system would also enable the system to be used for future personal space libraries of lesser extent than the typical public space libraries of the past.

Alternative Arrangements in the DDC of the Future

Second, the fact that there may be no *one best* system to classify knowledge does not mean that alternative arrangements of the DDC cannot be provided. Alternative arrangements have long been recognized and even occasionally suggested in the DDC as a way to overcome bias in the system or to accommodate important variations in its use. In the past, however, only specific alternatives have been suggested, and these have not always been promoted with enthusiasm for fear, one supposes, of undermining the standardization that the DDC has always supported.

Most of the problems associated with alternative arrangements have arisen because using them is so labor intensive. For example, should the classifier of a library collection wish to use one of the alternative facet sequences for the law schedules (a sequence not included in the classification copy), the classifier would have to allocate resources in time and labor to make the notational changes in the classification copy. An electronic version of the DDC holds some promise for reducing those costs, however, were it to include an automatic linking mechanism which would allow notations entered into it from a standard structure to be converted to alternative locations. The conversion could be made on the basis of alternatives that are either suggested by the system or are specially created for an individual library. Having such a linking mechanism would mean that if standard classification copy for law topics always used the facet sequence of Base Number—Branch of Law—Subordinate topic—Place, but a local library wished to use the sequence Base Number—Place—Branch of Law—Subordinate topic, the classification copy of the former could be converted to the latter with little more than a keystroke.

The other alternative arrangement, which involves moving from standard to tailored choices and collocations of categories, might be accommodated in a similar way, except that the mechanism for doing so would have to be more sophisticated. It would require something akin to a change template for moving an entire series of collocations from one place to another on the initiative of the DDC user. Thus, if a classifier wished to collocate a wide variety of information resources (or electronic links to them) related to jewelry, the template should allow the user to move the topical areas and subclasses of, say, gem metallurgy (553.8), gem mining (622.38), synthetic gems (666.88), gem carving (736.2), jewelry and gems in religion (the categories of individual religions from the 270s to 290s), and so on into a structure at, say, 739.27 (Jewelry) with links from their standard positions to the new locations.

New Approaches for Viewing and Managing the DDC of the Future

These suggestions are focused on making the DDC into a more malleable system than it is at present. However, to achieve this kind of malleability, we will have to view the structure of the DDC differently than we have in the past. We will have to see the entire system as a vast array of moveable or interchangeable facets of categories, a system that is perhaps best called an object relational database management system of categories. To turn the DDC into such a system will require, besides suitable linking and switching mechanisms, intense efforts to discover new hidden patterns of relationships among categories and sequences of categories so that these new patterns might be applied elsewhere than where they are typically applied at present. Efforts must also be made to discover hidden patterns of the use of these patterns of categories in the system. This might well be accomplished by systematically applying data-mining and knowledge discovery techniques to the scheme in its various capacities: as a database of categories; as a scheme to classify library collections; and as a tool to search for information.

The Purpose of the DDC of the Future

In the process of enhancing the DDC in this way, the foregoing work might also pave the way to our viewing the purpose of the system more broadly than we have viewed it in the past. Until now, the DDC has been viewed primarily as a document retrieval tool. Document retrieval is, of course, a legitimate use of classification and possibly will always be its central purpose. However, with a more finely tuned and malleable DDC, the system may also be useful for other purposes—for example, for teaching, learning, and memorizing knowledge categories, and for discovering relationships not previously imagined among diverse areas of categories and resources.

In this concluding section, I have raised a question about the future of the DDC. If the DDC continues on its present course of development—i.e., adopting the library classification theory and techniques that evolved in the first half of the twentieth century—is its future assured? Given the recent developments I have described that are related to the post-modern library and the post-modern age, it would seem unwise to follow the present course without making some changes. Simply following the present course would do little more than drive the system into another period of conflict and inconsistent change. The editors must heed the important changes occurring in the broader environment of which the DDC is a part. I believe that even modest efforts to accommodate such changes would ensure that the DDC has a future of enormous potentialities.

REFERENCE LIST

Editions of the DDC

1. 1876. *A Classification and Subject Index for Cataloging and Arranging the Books and Pamphlets of a Library* [Amherst, Mass.]. Facsimile reprint, Albany, N.Y.: Forest Press, 1976.

2. 1885. *Decimal Classification and Relativ Index for arranging, cataloging, and indexing* public and private libraries and for pamflets, clippings, scrap books, index rerums, etc. 2nd ed., rev. and greatly enl. Boston: Library Bureau.

3. 1888. *Decimal Classification and Relativ index for arranging, cataloging, and indexing* public and private libraries and for pamflets, clippings, notes, scrap books, index rerums, *etc.* 3rd ed., rev. and greatly enl. Boston: Library Bureau.

4. 1891. *Decimal Classification and Relativ Index for libraries, clippings, notes, etc.* 4th ed., rev. and enl. Boston: Library Bureau.

5. 1894. *Decimal Classification and Relativ Index for libraries, clippings, notes, etc.* 5th ed. Boston: Library Bureau.

6. 1899. *Decimal Classification and Relativ Index for libraries, clippings, notes, etc.* 6th ed. Boston: Library Bureau.

7. 1911. *Decimal Classification and Relativ Index for libraries, clippings, notes, etc.* Edition 7. Lake Placid Club, N.Y.: Forest Press.

8. 1913. *Decimal Classification and Relativ Index for libraries, clippings, notes, etc.* Edition 8. Lake Placid Club, N.Y.: Forest Press.

9. 1915. *Decimal Classification and Relativ Index for libraries, clippings, notes, etc.* [9th ed.] Lake Placid Club, N.Y.: Forest Press.

10. 1919. *Decimal Classification and Relativ Index for libraries, clippings, notes, etc.* Edition 10., rev. and enl. Lake Placid Club, N.Y.: Forest Press.

11. 1922. *Decimal Classification and Relativ Index for libraries and personal use, in arranjing* for immediate reference books, pamflets, clippings, pictures, manuscript notes and other *material.* Edition 11, rev. and enl. Lake Placid Club, N.Y.: Forest Press.

12. 1927. *Decimal Clasification and Relativ Index for libraries and personal use, in arranging* for immediate reference books, pamflets, clippings, pictures, manuscript notes and other *material.* Edition 12, rev. and enl. Semi-centennial ed. Lake Placid Club, N.Y.: Forest Press.

13. 1932. *Decimal Clasification and Relativ Index for libraries and personal use, in arranging* for immediate reference books, pamflets, clippings, pictures, manuscript notes and other *material.* Edition 13, revized and enlarjd by Dorkas Fellows, Editor, Myron Warren Getchell, Associate editor. Memorial ed. Lake Placid Club, N.Y.: Forest Press.

14. 1942. *Decimal Classification and Relativ Index.* Edition 14. Lake Placid Club, N. Y.: Forest Press.

15. 1951. *Decimal Classification.* Devised by Melvil Dewey. Standard (15th) edition. Lake Placid Club, N. Y.: Forest Press.

15rev. 1952. *Dewey Decimal Classification & Relative Index.* Devised by Melvil Dewey. Standard (15th) edition, revised. Lake Placid Club, N.Y.: Forest Press.

16. 1958. *Dewey Decimal Classification & Relative Index.* Devised by Melvil Dewey. 16th Edition. Lake Placid Club, N. Y.: Forest Press.

17. 1965. *Dewey Decimal Classification & Relative Index.* Devised by Melvil Dewey. Edition 17. Lake Placid Club, N. Y.: Forest Press, of Lake Placid Club Education Foundation.

18. 1971. *Dewey Decimal Classification & Relative Index.* Devised by Melvil Dewey. Edition 18. Lake Placid Club, N. Y.: Forest Press, of Lake Placid Club Education Foundation.

19. 1979. *Dewey Decimal Classification & Relative Index.* Devised by Melvil Dewey. Edition 19, edited under the direction of Benjamin A. Custer. Albany, N. Y.: Forest Press, a Division of Lake Placid Club Education Foundation.

19a (004-006). 1985. *DDC, Dewey Decimal Classification. Revision of Edition 19 004-006 Data processing and Computer science and changes in related disciplines.* Albany, N.Y.: Forest Press.

20. 1989. *Dewey Decimal Classification & Relative Index.* Devised by Melvil Dewey. Edition 20, edited by John P. Comaromi . . . [et al.]. Albany, N. Y.: Forest Press, a Division of OCLC Online computer Library Center.

21. 1997. *Dewey Decimal Classification & Relative Index.* Devised by Melvil Dewey. Edition 21, edited by Joan S. Mitchell . . . [et al.]. Albany, N. Y.: Forest Press, a Division of OCLC Online Computer Library Center.

Other Works

Anderson, Margaret. 1978. "Henry Evelyn Bliss (1870–1955)." In *Dictionary of American Library Biography*, 36–39. Littleton, Colo.: Libraries Unlimited.

Bliss, Henry E. 1929. *The Organization of Knowledge and the System of the Sciences*. New York: Holt.

————. 1933. *The Organization of Knowledge in Libraries*. New York: H. W. Wilson.

Brown, James D. 1914. *Subject Classification, with Tables, Indexes, etc. for the Subdivision of Subjects*. 2nd ed., rev. London: Grafton.

Cohen, I. Bernard. 1985. *Revolution in Science*. Cambridge, Mass.: The Belknap Press of Harvard University Press.

Coates, E. J. 1960. *Subject Catalogues: Headings and Structure*. London: The Library Association.

Comaromi, John. 1976. *The Eighteen Editions of the Dewey Decimal Classification*. Albany, N.Y.: Forest Press Division of the Lake Placid Education Foundation.

Cutter, Charles A. 1876a. "Library Catalogues." In *Public Libraries in the United States of America, their History, Condition, and Management. Special Report, Part 1*, 526–622. Washington, D.C.: Department of the Interior, Bureau of Education; [Distributed by the] Government Printing Office. [Reprinted in University of Illinois, Graduate School of Library Science, Monograph series no. 4.]

————. 1876b. *Rules for a Printed Dictionary Catalogue*. Part II of *Public Libraries in the United States of America, their History, Condition, and Management. Special Report*. Washington, D.C.: Department of the Interior, Bureau of Education; [Distributed by the] Government Printing Office.

————. 1897. "Reasons for Using the Expansive Classification in an International Bibliography." In Institut Internationale de Bibliographie. *Proceedings of the Second Conference held in London, July 13–16, 1897*, 196–99. Brussels: The Institut.

————. 1977a. "The Expansive Classification." In *Charles Ammi Cutter, Library Systematizer*. Edited by F. L. Miksa, 385–88. The Heritage of Librarianship series, no. 3. Littleton, Colo.: Libraries Unlimited. [Originally published in 1897.]

———. 1977b. "A Classification for the Natural Sciences." In *Charles Ammi Cutter, Library Systematizer.* Edited by F. L. Miksa, 258-59. The Heritage of Librarianship series, no. 3. Littleton, Colo.: Libraries Unlimited. [Originally published in 1880.]

D'Alembert, Jean Le Rond. 1963. *Preliminary Discourse to the Encyclopedia of Diderot.* Translated by R. N. Schwab, with the collaboration of W. E. Rex. Introduction and notes by R. N. Schwab. The Library of Liberal Arts. Indianapolis, Indiana: Bobbs- Merrill.

Dickson, David. 1988. *The New Politics of Science, with a new Preface.* Chicago: University of Chicago Press.

Dolby, R. G. A. 1979. "Classification of the Sciences: The Nineteenth Century Tradition." In *Classifications in their Social Context.* Edited by R. F. Ellen and D. Reason, 167–93. New York: Academic Press.

Eaton, Thelma. 1952. "Dewey Re-examined." *Library Journal* 77 (May 1, 1952): 745–51.

Flanzraich, Gerri. 1993. "The Library Bureau and Office Technology." *Libraries and Culture* 28, no. 4 (fall): 403–29.

Flint, Robert. 1904. *Philosophy as Scientia Scientiarum and A History of Classifications of the Sciences.* Edinburgh: William Blackwood.

Graziano, Eugene E. 1959. "Hegel's Philosophy as Basis for the Decimal Classification Schedule." *Libri* 9: 45–52.

Holton, Gerald. 1996. *Einstein, History, and other Passions: The Rebellion against Science at the End of the Twentieth Century.* Reading, Mass.: Addison-Wesley.

Huxley, Aldous. 1932. *Brave New World, a Novel.* London: Chatto and Windus.

Jevons, William S. 1877. *The Principles of Science: A Treatise on Logic and Scientific Method.* 2nd ed., rev. London, New York: MacMillan.

LaMontagne, Leo E. 1961. *American Library Classification, with Special Reference to the Library of Congress.* Hamden, Conn.: Shoe String Press.

Learned, William S. 1924. *The American Public Library and the Diffusion of Knowledge.* New York: Harcourt, Brace.

Machlup, Fritz. 1980. *Knowledge: Its Creation, Distribution, and Economic Significance.* Vol. 1, *Knowledge and Knowledge Production.* Princeton, N.J.: Princeton University Press.

————. 1982. *Knowledge: Its Creation, Distribution, and Economic Signifi-cance.* Vol. 2, *The Branches of Learning.* Princeton, N.J.: Princeton University Press.

Mayr, Ernst. 1982. *The Growth of Biological Thought: Diversity, Evolution, and Inheritance.* Cambridge, Mass.: The Belknap Press of Harvard University Press.

Miksa, Francis L. 1974. "Charles Ammi Cutter: Nineteenth-Century Systemizer of Libraries." Ph.D. dissertation, University of Chicago.

————. 1983a. *The Subject in the Dictionary Catalog from Cutter to the Present.* Chicago: American Library Association.

————. 1983b. "Melvil Dewey and the Corporate Ideal." In *Melvil Dewey: The Man and the Classification.* Edited by G. Stevenson and J. Kramer-Greene, 49–100. Albany, NY: Forest Press.

————. 1984. *The Development of Classification at the Library of Congress.* Occasional Papers, no. 164. Champaign, Ill.: University of Illinois, Graduate School of Library and Information Science.

————. 1985. "Machlup's Categories of Knowledge as a Framework for Viewing Library and Information Science History." *Journal of Library History* 20, no. 2 (spring 1985): 157– 72.

————. 1986. "Melvil Dewey: The Professional Educator and His Heirs." *Library Trends* 34 (winter): 359-81.

————. 1987. *Research Patterns and Research Libraries.* Dublin, Ohio: OCLC, Online Computer Library Center, Inc.

————. 1988. "The Columbia School of Library Economy, 1887–1888." *Librar-ies and Culture* 23, no. 3 (summer): 249 –80.

————. 1994. "Classification." In *Encyclopedia of Library History.* Edited by W. A. Wiegand and D. G. Davis, Jr., 145–53. New York: Garland.

————. 1996. "The Cultural Legacy of the 'Modern Library' for the Future." *Journal of Education for Library and Information Science* 37, no. 2 (spring): 100–19.

Ortega y Gasset, Jose. 1975. "The Mission of the Librarian." In *Of, By, and For Librarians, Second Series.* Edited by John D. Marshall, 190–213. Hamden, Conn.: Shoe String Press. [Originally given as an address to the International Congress of Bibliographers and Librarians, Paris, 1934, and published in Spanish.]

Pollard, A.F.C. 1925. "The Decimal Classification of the Institut International de Bibliographie and Its Importance As a Key to the World's Literature." *ASLIB Proceedings* 2: 37–47.

Random House Dictionary of the English Language. 1987. Second edition, unabridged. New York: Random House.

Ranganathan, S. R. 1931. *The Five Laws of Library Science.* Madras Library Association, Publication Series, no. 2. Madras: Madras Library Association; London: Edward Goldston.

———. 1933. *Colon Classification.* Madras Library Association, Publication Series, no. 3. Madras: Madras Library Association; London: Edward Goldston.

———. 1934. *Classified Library Code.* Madras Library Association, Publication Series, no. 4. Madras: Madras Library Association; London: Edward Goldston.

———. 1935. *Library Administration.* Madras Library Association, Publication Series, no. 5. Madras: Madras Library Association; London: Edward Goldston.

———. 1937. *Prolegomena to Library Classification.* Madras Library Association, Publication Series, no. 6. Madras: Madras Library Association; London: Edward Goldston.

———. 1938. *Theory of Library Catalogue.* Madras Library Association, Publication Series, no. 7. Madras: Madras Library Association; London: Edward Goldston.

———. 1939. *Colon Classification.* 2nd ed., rev. Madras Library Association, Publication Series, no. 8. Madras: Madras Library Association; London: Edward Goldston.

———. 1940. *Reference Service and Bibliography.* Madras Library Association, Publication Series, no. 9. Madras: Madras Library Association; London: Edward Goldston.

———. 1942. *School and College Libraries.* Madras Library Association, Publication Series, no. 11. Madras: Madras Library Association; London: Edward Goldston.

———. 1944. *Library Classification: Fundamentals & Procedure.* With 1008 graded examples and exercises. Madras Library Association, Publication Series, no. 12. Madras: Madras Library Association; London: Edward Goldston.

———. 1945. *Elements of Library Classification.* Based on lectures delivered at the University of Bombay in December 1944. Poona City: N. K. Publishing House.

————. 1950a. *Colon Classification.* 3rd ed. Madras Library Association, Publication Series, no. 16. Madras: Madras Library Association.

————. 1950b. *Library Tour 1948, Europe and America, Impressions and Reflections.* Indian Library Association, English Series. Delhi: Indian Library Association; London: G. Blunt.

————. 1951a. *Classification and Communication.* Delhi: University of Delhi.

————. 1951b. *Philosophy of Library Classification.* Library Research Monographs, Vol. 2. Copenhagen: Ejnar Munksgaard.

————. 1952. *Colon Classification.* 4th ed. Madras Library Association, Publication Series, no. 19. Madras: Madras Library Association; London: G. Blunt.

————. 1957a. *Colon Classification.* 5th ed. Madras Library Association, Publication Series, no. 22. Madras: Madras Library Association.

————. 1957b. *Prolegomena to Library Classification.* 2nd ed. With a preface by W. C. Berwick Sayers. London: The Library Association.

————. 1960. *Colon Classification.* 6th ed., rev. completely. Madras Library Association, Publication Series, no. 26. Bombay, New York: Asia Publishing House.

————. 1962. *Elements of Library Classification.* Based on lectures delivered at the University of Bombay in December 1944 and in the Schools of Librarianship in Great Britain in December 1956. Bombay; London: Asia Publishing House.

————. 1963. *Colon Classification.* 6th ed., reprinted with amendments. Madras Library Association, Publication Series, no. 26. Bombay, New York: Asia Publishing House.

————. 1967. *Prolegomena to Library Classification.* 3rd ed. Assisted by M. A. Gopinath. Bangalore: Sarada Ranganathan Endowment for Library Science.

————. 1987. *Colon Classification.* Edition 7 (Basic and Depth Version), revised and edited by M. A. Gopinath. Bangalore: Sarada Ranganathan Endowment for Library Science.

Rayward, W. Boyd. 1975. *The Universe of Information: The Work of Paul Otlet for Documentation and International Organisation.* FID 520. Moscow: All-Union Institute for Scientific and Technical Information—VINITI.

Richardson, Ernest C. 1930. *Classification: Theoretical and Practical.* 3rd ed. New York: H. W. Wilson. [When this work was first published, in 1901, it consisted of lectures delivered at the New York State Library School at Albany, N.Y. It was reprinted in 1912 without any changes. Some appendices were added to the 1930 reprint cited here, but essentially, this reprint contains the same materials on the topic of classification that Richardson had presented in 1901.]

Sayers, W. C. Berwick. 1912. *The Grammar of Classification*. The Library Assistants' Association Series, no. 1. Croyden: Central Library. [A briefer first edition had been published in 1908, a third in 1924, and a fourth in 1935.]

————. 1915–1916. *Canons of Classification Applied to "The Subject", "The Expansive", "The Decimal", and "The Library of Congress" Classifications: A Study in Bibliographical Classification Method*. London: Grafton; White Plains, N.Y.: H. W. Wilson.

————. 1918. *An Introduction to Library Classification, with Readings, Questions and Examination Papers*. London: Grafton; New York: H. W. Wilson. [This work went through nine editions by 1954.]

————. 1926. *A Manual of Classification for Librarians & Bibliographers*. London: Grafton. [Sayers issued a second edition of this work in 1944 and a third in 1955. Subsequently, the fourth and fifth editions of this work were revised by Arthur Maltby. *A New Manual of Classification* by Rita Marcella and Robert Newton continued the series in 1994.]

Scott, Edith. 1960. "Dewey Reviews: From the U.S." *Library Resources and Technical Services* 4, no. 1 (winter): 14–23.

————. 1970. "J. C. M. Hanson and His Contribution to Twentieth Century Cataloging." Ph.D. dissertation. University of Chicago.

Shamurin, E. I. 1955–59. *Ocherki Po Istorii Bibliotechno-Bibliograficheskaia Klassifikatsii*. Moscow: Vsesoiuznaia Knizhnaia Palata.

Shera, Jesse H. and Egan, Margaret. 1953. "Introduction." In Bradford, Samuel C. *Documentation*. 2nd ed. London: Lockwood.

Spengler, Oswald. 1926. *The Decline of the West*. New York: Knopf. [Translation of *Der Untergang des Abendlandes* (1918)]

Steinberg, Steve G. 1996. "Seek and Ye Shall Find (Maybe): Indexing the Web." *Wired* 4, no. 5 (May): 108–14, 172, 174, 176, 178, 180, 182.

Vickery, B. C. 1975. "Historical Aspects of the Classification of Science." Appendix A in *Classification and Indexing in Science*, 147–80. 3rd ed. London: Butterworths.

Wiegand, Wayne A. 1986. *The Politics of an Emerging Profession: The American Library Association, 1876–1917.* Contributions in Librarianship and Information Science, no. 56. New York: Greenwood Press.

————. 1996a. *Irrepressible Reformer: A Biography of Melvil Dewey.* Chicago: American Library Association.

————. 1996b. "Origins of the Dewey Decimal Classification System: Another Look." Manuscript in preparation for publication.

The DDC, the Universe of Knowledge, and the Post-Modern Library was designed and composed in Times Roman typeface by Lisa Hanifan of Albany, New York. The book was printed by **Integrated Book Technology of Troy, New York.**